Stable Isotopes to Trace Migratory Birds
and to Identify Harmful Diseases

G.J. Viljoen • A.G. Luckins • I. Naletoski

Stable Isotopes to Trace Migratory Birds and to Identify Harmful Diseases

An Introductory Guide

Front Photo
FAO Mediabase; Credit ©FAO/Giulio Napolitano/FAO

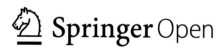

G.J. Viljoen
Joint FAO/IAEA Division
Animal Production and Health Section
Vienna, Austria

A.G. Luckins
Schiehallion
Blairgowrie Schotland, UK

I. Naletoski
Joint FAO/IAEA Division
Animal Production and Health Section
Vienna, Austria

ISBN 978-3-319-28297-8   ISBN 978-3-319-28298-5   (eBook)
DOI 10.1007/978-3-319-28298-5

Library of Congress Control Number: 2016933681

Printed on acid-free paper

This Springer imprint is published by Springer Nature
The registered company is Springer International Publishing AG Switzerland

Foreword

Highly pathogenic avian influenza (HPAI) caused by the H5N1 virus is a dangerous transboundary animal disease (TAD) that seriously impacts on poultry production, especially on smallholder farmers and commercial enterprises in Member States (MS) in Asia and Africa. With the added danger that this disease can become zoonotic, i.e. it has the potential to transfer to humans, it is essential that every effort is made to understand its epidemiology, particularly those factors that determine the dissemination of the virus across countries and continents. One area in which there is an urgent need for clarity is the role that migratory wild birds might play in the spread of HPAI to domestic poultry. This is a matter of international importance since millions of wild birds fly from areas where HPAI is endemic to all corners of the globe, potentially spreading the disease wherever they land.

MS are acutely aware of the need to develop technologies that allow disease detection before clinical signs of infection became apparent. Under the aegis of the project "Reducing risk from transboundary animal diseases (TADs) and those of zoonotic importance", the IAEA, through the Animal Production and Health Section of the Joint FAO/IAEA Division, has supported MS in acquiring sensitive, specific and rapid nuclear, nuclear-related and nuclear-associated diagnostic tests for HPAI. Now that MS laboratories have the required technologies to develop appropriate surveillance strategies, it will be possible to enlarge the scope of studies on HPAI to include verification of the role of the wild bird hosts. Consequently, the Animal Production and Health Section (APHS) has initiated a new Coordinated Research Project (CRP) that uses innovative technologies to sample wild birds and their habitats to detect virus and also introduce stable isotope analysis (SIA) of bird tissues to determine their place of origin.

Migration patterns of wild birds have been studied for many years using a variety of techniques, including recent various electronic tracking methods. However, more effective and cheaper options will be required if specific information is needed on a particular bird species, especially infected wild birds. The envisaged approach is to determine migrant origins by determining the stable isotope "signature" of the bird and match it to the most likely environmental habitat. Stable isotopes pose no health or environmental risks since they are not radioactive and are present throughout the environment. However only a small number of them, namely, H, O, C, N and S, will be useful for tracking the origin of wild birds. The technologies used to infer migratory connectivity in migrant birds and for detection have become increasingly sophisticated, as have the models for identifying habitats. It is now felt that such technologies are mature enough to inform scientists engaged in studies of HPAI on so far unanswered questions concerning the biology and epidemiology of HPAI in wild birds.

The purpose of this publication is to make scientists, particularly those involved in IAEA CRP and TC projects, aware of the potential of SIA by providing an introduction of their application and technical requirements needed to facilitate studies.

Animal Production and Health Section   Gerrit Viljoen
Joint FAO/IAEA Division for Nuclear Applications
in Food and Agriculture, Department of Nuclear
Sciences and Applications
International Atomic Energy Agency
Vienna International Centre
Vienna, Austria

Acknowledgements

This guide was developed with a substantial support of the team involved in the development and implementation of the IAEA CRP: "Use of Stable Isotopes to Trace Bird Migrations and Molecular Nuclear Techniques to Investigate the Epidemiology and Ecology of the Highly Pathogenic Avian Influenza" (Code: D32030).

A second edition of *Stable Isotopes to Trace Migratory Birds and to Identify Harmful Diseases – An Introductory Guide* is in preparation and will supersede the original published in 2016.

Reference

Tracking Animal Migration with Stable Isotopes, Edited by: Keith A. Hobson, Leonard I. Wassenaar, First Edition 2008 (ISBN: 978-0-12-373867-7, ISSN:1936-7961).

Contents

List of Acronyms

Acronym	Interpretation
APHS	Animal Production and Health Section of the Joint FAO/IAEA Division
BWB	Bowhead Whale Baleen
C	Carbon
CDT	Canyon Diablo Troilite
CF-IRMS	Continuous flow isotope ratio mass spectrometer
CFS	Chicken feather standard
CHS	Cow hoof standard
CRP	Coordinated Research Project
D	Deuterium
DNA	Deoxyribonucleic acid
ELISA	Enzyme-linked immunosorbent assay
EU	European Union
FAO	Food and Agricultural Organization of the United Nations
GIS	Global information system
GLM	General linear model
GNIP	Global Network for Isotopes in Precipitation
GPS	Global Positioning System
GS	Growing season
H	Hydrogen
HDPE	High-density polyethylene
HPAI	Highly pathogenic avian influenza
IAEA	International Atomic Energy Agency
IRMS	Isotope ratio mass spectrometer(spectrometry)
LPAI	Low pathogenic avian influenza
MA	Mean annual
MS	Member states
N	Nitrogen
O	Oxygen
OPIC	Online Isotopes in Precipitation Calculator

PDB	Pee Dee Formation
QA	Quality assurance
RBC	Red blood cells
Rs	Ratio of the heavy and the light isotope of the sample
Rstd	Ratio of the heavy and the light isotope of the standard
S	Sulphur
SIA	Stable isotope analysis
SLAP	Standard light Antarctic precipitation
SMOW	Standard Mean Ocean Water
TAD	Transboundary animal disease
TC	Technical cooperation
VPDB	International Atomic Energy Agency – Pee Dee Formation
VSMOW	International Atomic Energy Agency – Standard Mean Ocean Water
WMO	World Meteorological Organization

The original version of this book was revised: The acknowledgements have been updated to reflect the contributions of K.A. Hobson and L.I. Wassenaar to the project. The correction to this chapter is available at https://doi.org/10.1007/978-3-030-25101-7_4

Chapter 1
General Introduction

1.1 Background Analysis

Birds are amongst the few terrestrial vertebrates that share with humans the pecu-liarity of travelling in a few hours across national and intercontinental borders (Hobson 2002). The record for distance covered in a single year belongs to the arctic tern, which travels over 50,000 km between Antarctica and northern Scandinavia. Overall, billions of birds travel between continents twice a year in only a few weeks. Migration is critical in the life cycle of a bird, and without this annual journey many birds would not be able to raise their young. More than 5000 species of birds manage annual round-trip migrations of thousands of miles, often returning to the exact same nesting and wintering locations from year to year. Birds migrate to find the richest, most abundant food sources that will provide adequate energy to nurture young birds. If no birds migrated, competition for adequate food during breeding seasons would be fierce and many birds would starve. Instead, birds have evolved different migration patterns, times and routes to give themselves and their offspring the greatest chance of survival. Birds gauge the changing of the seasons based on light level from the angle of the sun in the sky and the amount of daily light. When the timing is right for their migrating needs, they will begin their journey. Several minor factors can affect the precise day any bird species begins its migration, including available food supplies, poor weather or storms and air tem-peratures and wind patterns. While these factors may affect migration by a day or two, most bird species follow precise migration calendars. Those calendars vary widely for different species, however, and while autumn and spring are peak migra-tion periods when many birds are on the move, migration is actually an ongoing process and at any time of the year, there are always birds at some stage of their journeys. The distance the birds must fly, the length of time it takes to mate and produce a healthy brood, the amount of parental nurturing young birds receive and the location of birds' breeding and wintering grounds all affect when any one spe-cies migrates to stay alive.

© IAEA 2016
G.J. Viljoen et al., *Stable Isotopes to Trace Migratory Birds and to Identify Harmful Diseases*, DOI 10.1007/978-3-319-28298-5_1

During these yearly migrations, birds have the potential of dispersing microorganisms that can be dangerous for human as well as animal health. For instance, birds are believed to be responsible for the wide geographic distribution of various pathogens, including viruses [e.g., West Nile, Sindbis, Highly Pathogenic Avian Influenza (HPAI), Newcastle disease], bacteria (e.g., *Borrelia, Mycobacterium, Salmonella*), and protozoa (e.g., *Cryptosporidium*). An insight into the ecology of bird populations is necessary to understand fully the epidemiology of bird-associated emerging diseases. Furthermore, data about avian movements might be used to improve disease surveillance or to adapt preventive measures. However, the links between bird ecology and livestock and human disease have yet to be completely understood so there is a need to increase knowledge of avian migration patterns and infectious diseases to help predict future outbreaks of emerging diseases.

Animal influenza viruses threaten animal health, livestock productivity and food security in poor countries, but they can also evolve into dangerous human pathogens. This has been seen with the emergence of HPAI. Its main impact has been on domesticated poultry, with over 300 million birds killed or destroyed, but of considerable public health concern is the transmission of the virus from birds to humans, resulting in over 500 occurrences of disease in which over half of the infected individuals have died.

The threat from animal influenza viruses makes it essential for animal health professionals to take the lead in detecting and monitoring the occurrence of the viruses and sharing the information with the international community. Since avian influenza appears to be associated with migratory bird movements (over 1000 reported AI outbreaks since 2006, involving 25 species of wild birds in EU alone) surveillance would need to focus on detection of HPAI in both wild birds as well as domesticated poultry and it will also be necessary to establish the migratory pathways of wild birds to increase the capacity to assess their risk in spreading the virus.

In March 2010 the International Scientific Task Force on Avian Influenza and Wild Birds, led by FAO and the United Nations Environment Programme – Convention on Migratory Species, reported that waning attention to HPAI was reducing opportunities for surveillance and research, thereby affecting efforts to understand the epidemiology of the disease. The disease continues to be a major problem in Egypt and parts of Asia and outbreaks have occurred in poultry in Romania and in wild birds in Russia, China and Mongolia (Fig. 1.1). One of the most important issues that need addressing is the surveillance of wild bird populations to improve understanding of the role that they play in the dissemination of infection. Although various methods have been used to track the migration of birds there is increasing interest in utilizing methodologies that would enable tracing of migratory movement based on the birds' stable isotope signatures. A small number of stable isotopes are involved in important biological and ecological processes and there is a strong correlation between levels of these isotopes in the environment and the concentration of the same isotopes in avian tissues. Of most interest are stable hydrogen and oxygen ratios in tissues that accurately reflect those in lakes, rivers, oceans and in groundwater, along the wild bird flyways. Using stable isotopes to characterize a population involves examining the isotopic signatures of a few

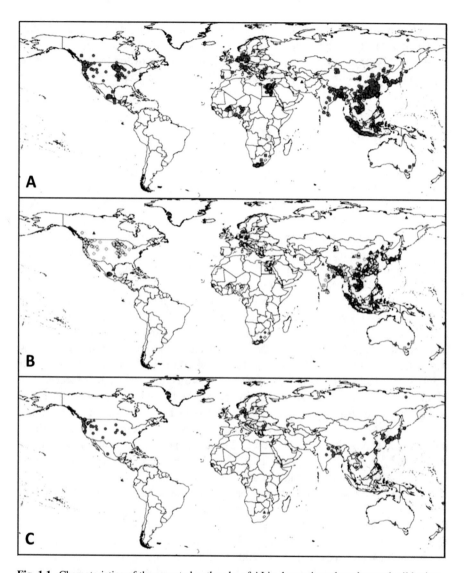

Fig. 1.1 Characteristics of the reported outbreaks of AI in domesticated poultry and wild migratory birds in the period from 1 January 2011 until 13 May 2015 (Source: FAO 2015).
Map A symbols: *red circle* [●]=H5 AI; *blue square* [■]=H7 AI; *green triangle* [▲]=H9 AI; *grey star* [★];
Map B symbols: AI outbreaks reported during 2011 (*grey star* [★]), 2012 (*green square* [■]), 2013 (*blue square* [◆]), 2014 (*red triangle* [▲]) and 2015 (*yellow circle* [○]);
Map C symbols: AI outbreaks in wild (*red circle* [●]) and captive birds (*blue triangle* [▲])

individuals that are representative of the entire population. The hydrogen and oxygen isotope composition of environmental water varies spatially across the globe and because it is a constituent of many biosynthetic pathways, the isotopes' presence is relayed to animal tissues, providing the means to link data on groundwater isoscapes with isotope levels in biological tissues such as feathers. This isotope data would reveal migration patterns and enable identification of the breeding areas of birds sampled at non-breeding grounds and disease outbreak sites.

1.2 Migratory Birds and HPAI

To date, only a small number of migratory birds have been tracked by satellite transmitters to establish links with disease outbreaks in domestic poultry, with little evidence so far of direct correlation with HPAI. Furthermore, surveillance of 750,000 "healthy" wild birds has not revealed many infected individuals. Circumstantial evidence suggesting that the spread of the H5N1virus to new areas can be facilitated by migratory wild birds has come from studies on ducks in Asia marked with satellite transmitters (Yamaguchi et al. 2010) that were tracked during an outbreak of highly pathogenic H5N1 avian influenza virus.

The satellite transmitters were attached to northern pintail ducks several months before the H5N1 virus was discovered in dead and dying whooper swans at wetlands in Japan. Twelve percent of marked pintails used the same wetlands as infected swans and the pintail ducks were present at those sites on dates the virus was discovered in swans. During the first week after they become infected with H5N1 virus, ducks such as pintails can shed the virus orally or in their faeces, contributing to the virus' spread. Some of the marked pintails migrated 700 miles within 4 days of leaving the outbreak sites; marked pintails ultimately migrated more than 2000 miles to nesting areas in eastern Russia. The discovery that northern pintails made long-distance migrations during the period when an infected duck would likely shed the virus offers insight into how H5N1 could be spread by wild birds across large areas (Yamaguchi et al. 2010).

In another study, the movements of bar-headed geese marked with GPS satellite transmitters at Qinghai Lake, China were traced in relation to virus outbreaks and disease risk factors. A previously undocumented migratory pathway between Qinghai Lake and the Lhasa Valley of Tibet where 93 % of the 29 marked geese overwintered was discovered. From 2003 to 2009, 16 outbreaks in poultry or wild birds were confirmed on the Qinghai-Tibet Plateau, and the majority were located within the migratory pathway of the geese. Spatial and temporal concordance between goose movements and three potential H5N1 virus sources (poultry farms, a captive bar-headed goose facility, and H5N1 outbreak locations) indicated that ample opportunities existed for virus spill over and infection of migratory geese on the wintering grounds. Their potential as vectors of H5N1 was supported by rapid migration movements of some geese and genetic relatedness of H5N1 virus isolated from geese in Tibet and Qinghai Lake. This study was the first to compare phylogenetics of the virus with spatial ecology of its host, and the combined results suggest that the geese play a role in the spread of H5N1 in this region (Prosser et al. 2006; Zhou et al. 2011).

Defining the migratory behaviour of animals is fundamental to understanding their evolution and life history; much of the effort to study migration in birds, has been driven by the needs for conservation in changing habitats. The advances in understanding of migration, particularly in North America and Europe have been aided considerably by stable isotope techniques. The methods have not yet been applied to study the migratory birds and the dissemination of HPAI, but this is likely to be remedied in the near future. This manual provides background information on Stable Isotope Analysis (SIA) for scientists investigating HPAI from Member States engaged in Collaborative Research Projects or Technical Contract Projects and wish to understand how the disease is disseminated to provide information that would enable better risk assessment and more effective control and prevention measures.

In order to follow the global distribution of animal diseases, as well as to enable rapid information sharing, FAO has established an Animal Disease Information System (EMPRES-i), (FAO 2015) which enables for filtering disease records according to different predefined criteria. Specifically for AI, record-sets can be generated, which enable for correlation according to the species of origin of the outbreak (domesticated or wild birds), period of the observation/reporting of the outbreak and the location of the outbreak. The data on the evolution of the AI globally, obtained from EMPRES-i system over the period of 5 years (January 2011–May 2015) is shown on Fig. 1.1 and Table 1.1.

1.3 Using SIA to Understand the Dissemination of HPAI – The Way Ahead!

Most elements consist of one or more stable isotopes – elements having the same number of protons, but differing in the numbers of neutrons. Stable isotopes are those isotopes of an element that do not decay through radioactive processes over time. For instance, the element carbon (C) exists as two stable isotopes, ^{12}C and ^{13}C, and the element hydrogen (H) exist as two stable isotopes, ^{1}H and ^{2}H.

Stable isotope contents are expressed in 'delta' notation as δ values in parts per thousand (‰), where:

$$\delta R‰ = (Rs / RStd - 1) \times 1000$$

and Rs and RStd are the ratios of the heavy to light isotope (e.g. $^{13}C/^{12}C$) in the sample and the standard, respectively. The stable isotope ratios of hydrogen, carbon, nitrogen, oxygen and sulphur are denoted in delta notation as $\delta^{2}H$, $\delta^{13}C$, $\delta^{15}N$, $\delta^{18}O$ and $\delta^{34}S$, respectively. R values have been carefully measured for internationally recognized standards. The standard used for both H and O is Standard Mean Ocean Water (SMOW), where $(^{2}H/^{1}H)$ standard is 0.0001558 and $^{18}O/^{16}O$ is 0.0020052. The original SMOW standard is no longer available and has been replaced by a new International Atomic Energy Agency (IAEA) standard, VSMOW. The international

Table 1.1 Quantitative information on the reported outbreaks of AI in domesticated poultry and wild migratory birds in the period from 1 January 2011 until 13 May 2015 (Source: FAO 2015)

Species category	Pathogenicity (HP or LP) – H subtype of the AIV	2011	2012	2013	2014	2015	Sub total
Captive		Number of outbreaks/number of cases affected during the utbreak					
	HP-H5	1/1			1/	8/5	10/6
	LP-H5		2/8				2/8
	LP-H7			1/4			1/4
	Total captive	1/1	2/8	1/4	1/	8/5	13/18
Domestic	HP-H5	1927/763,007	551/666,346	489/306,795	891/520,219	1614/2,359,345	5472/4,615,712
	HP-H7	1/10	48/2,699,685	86/780,143	4/3	3/139	142/3,479,980
	HP-NoInfo	1/10					1/10
	LP-H5	2/40	24/1584	37/6071	25/8179	24/15,526	112/31,400
	LP-H7	24/4369	1/	40/7186	34/3039	6/2373	105/16,967
	LP-H9	39/45,003	69/3	35/6	8/	4/	155/45,012
	LP-NoInfo	6/26		1/1		2/	9/27
	NoInfo-NoInfo			34/2519			34/2519
	Total domestic	2000/812,465	693/3,367,618	722/1,102,721	962/531,440	1653/2,377,383	6030/8,191,627

Species category	Pathogenicity (HP or LP) – H subtype of the AIV	2011	2012	2013	2014	2015	Sub total
Wild	HP-H5	63/101	25/1183	4/2	37/35	70/217	199/1538
	HP-H7		2/11	1/5		1/8	4/24
	HP-NoInfo					1/	1/
	LP-H10				1/		1/
	LP-H5	1/			2/		3/
	LP-H7			1/	1/		2/
	LP-H9				1/		1/
	Total wild	64/101	27/1194	6/7	42/35	72/225	211/1562
	Grand total	2126/812,567	7557/3,368,820	921/1,102,732	1376/531,475	2066/2,377,613	7246/8,193,207
NoInfo	HP-H5	59/	34/	38/	39/	147/	317/
	HP-H7		1/				1/
	LP-H10			1/	2/		3/
	LP-H7			152/	329/	185/	666/
	LP-H9	2/		1/	1/	1/	5/
	Total NoInfo	61/	35/	192/	371/	333/	992/

Confirmed AI outbreaks with no record on the total number of cases and the species affected (Source: FAO 2015) carbon standard is the PDB, where (^{13}C/^{12}C) is 0.0112372 and is based on a belemnite from the Pee Dee Formation. As with SMOW, the original PDB standard is no longer available, but the IAEA provides Vienna-PDB with a similar R value. Atmospheric nitrogen is the internationally recognized standard with an R value (^{15}N/^{14}N) of 0.0036765. Lastly, the internationally recognized standard for sulfur is CDT, the Canyon Diablo Troilite, with a value (^{34}S/^{32}S) of 0.0450045. Typically, during most stable isotope analyses, investigators would not use IAEA standards on a routine basis. Instead, laboratories establish secondary reference materials to use each day that are traceable to IAEA standards and that bracket the range of isotope ratio values anticipated for the samples.

Although ecologists refer to "isotope signatures or isotope fingerprints" the values obtained for a sample do not provide a unique fingerprint but a distinctive profile; a more neutral notation is isotope value, rather than signatures.

The δ notation is derived as follows:

$$\delta H\,(\%o) = \left(\frac{Isotopic\ ratio\ sample}{Isotopic\ ratio\ standard} - 1 \right) \times 1000$$

The right side of the equation is the measure of the light to heavy isotope (^{2}H/^{1}H) and that ratio is multiplied by 1000 transform the values into whole numbers. The isotope reference points were established many years ago and results reported can be negative or positive (‰) relative to the accepted international standard. Thus, a δH value of +150 (‰) means the sample has 150 parts per thousand (15 %) more deuterium in it than the standard; while if the value were negative it would be 150 times less deuterium than the standard. Primary reference material is limited in quantity so laboratories tend to use local standards calibrated against a reference standard. The primary isotopic reference standard for hydrogen and oxygen is Vienna Standard Mean Ocean Water (VSMOW).

It will be necessary to develop suitable procedures for monitoring wild birds to detect HPAI, determine their origin and migratory pathway to estimate the duration of their stay at the place of capture. This will require the application of detailed standard operating procedures for feather and tissue collection, data recording, sample storage, and designation of a laboratory to carry out the SIA. Since tissue samples could come from birds suspected of carrying avian influenza virus it will be necessary to install safeguards to prevent possible infection. Soft tissues could be freeze-dried and both these and keratinaceous samples heated to 100 °C for 20 min to destroy any viruses. Alternative methods would be to irradiate specimens at a central location before making them available for study. Various innovative methods for virus detection should also be considered, including environmental sampling to detect H5N1 virus in water, or faeces (Khalenkov et al. 2008; Cheung et al. 2009; Dovas et al. 2010) in locations where wild birds congregate. Other procedures that will be useful could include identification of bird species using DNA barcoding from feather or faeces (Lee et al. 2010) to provide better links to identifying those birds that are truly implicated in dissemination of HPAI.

There are already available data on variation in stable isotope ratios across the globe and how this is reflected in tissue samples from birds inhabiting different regions. It might be necessary however to collect environmental samples from different locations if there is not sufficient data already available. Collection of wild birds is a specialized procedure and it will be necessary to liaise with wildlife groups skilled in this task, who are also able to provide basic data on the ecology of different bird species. The FAO Wildlife Unit can provide inputs in this area through its links with various wildlife organizations in order to carry out sampling with greater efficiency.

Consideration needs to be given to which isotopes to analyse. While stable hydrogen can be used to obtain information on the wider geographical location, further details might be inferred on the local habitat by analysing carbon, nitrogen and sulphur. Samples from keratinaceous tissues e.g. feathers and claws, will provide an isotope ratio of the place when they were grown. In blood, stable isotope ratios could be used to determine timing of arrival on breeding or wintering grounds. Isotope ratios in the blood relate to those in the current environment hence a difference between isotope ratios in blood and the environment would indicate newly arrived birds compared with birds that had been staying in a particular location for some time. If infected birds were to be found it should be possible to establish where the disease was acquired – at the site or in a previous location. By understanding migratory movements it will be possible to predict risk and derive models to show the spread of HPAI and identify areas that pose a significant potential for being locations where the dissemination of HPAI is greater due to the particular congregation of wild birds, H5N1 virus and domestic poultry.

Chapter 2
Animal Migration Tracking Methods

2.1 Extrinsic Markers

The migration of birds has been studied for many years relying mostly on extrinsic passive markers attached to individual animals at the point of capture, with the expectation that a proportion of the marked individuals will then be identified in another location at a different point in time. Over the past 100 years, the most widespread approach has been through the application of markers such as leg bands, neck collars, or dyes. Many millions of birds have been tagged in this way but although this method has provided insights into migration, for the vast majority of bird species examined the recovery rate is low. An alternative to these simple devices is to use miniature transmitting devices – radio transmitters, radar and satellite tracking – that serve as active markers and are small enough (<0.5 g) to be attached to even small birds or mammals. The location of the marked animal can be inferred by tracing the individual using a receiver, or by triangulation using several receivers. Since the devices are miniaturized their range and battery life are restricted and they can provide information over only a few kilometers. Radar technology has also made useful contributions to studies on migration since it can provide information on animal movements over considerable distances, but since radar installations are fixed it is not possible to trace movements over the whole spectrum of migration routes used by birds. The most significant advances in tracking migratory animals have come from the use of satellite transmitters that allow highly accurate positioning of individual animals (Hiroyoshi and Pierre 2005; Whitworth et al. 2007; http://www.fao.org/avianflu/en/wildlife/sat_telemetry.htm). Much of the globe is covered by satellites so that animals can be monitored over thousands of kilometers. The technique can only be used on relatively large animals as the weight of the smallest transmitters is approximately 10 g, restricting their use to an animal weighing about 250 g, thereby excluding 80 % of the world's birds and 70 % of the mammals. With the exception of satellite transmitters all extrinsic markers require that individuals be recaptured, re-sighted or move within a detector's range at some time after initial

---

The original version of this chapter was revised: The chapter was inadvertently published with materials that were reproduced or modified without permission for the table headings in Tables 2.1–2.3 and 2.5 which has been updated now and also fix the citation for author Wunder and Norris (2008a and 2008b) in Table 2.5 and (2008b) on page 31. The correction to this chapter is available at https://doi.org/10.1007/978-3-319-28298-5_4

© IAEA 2016, corrected publication 2022
G.J. Viljoen et al., *Stable Isotopes to Trace Migratory Birds and to Identify Harmful Diseases*, DOI 10.1007/978-3-319-28298-5_2

marking. The probability of recapture depends on the number of observers, the regions and habitats and the chances of success are low. In addition, extrinsic methods tend to be biased towards regions with a high likelihood of mark-recapture (Hobson et al. 2004). A fundamental flaw in the use of an extrinsic marker is that it provides information only on the marked individuals. Geolocators and satellite tracking rely on small sample sizes and the devices may affect the behavior of the marked bird (Stutchbury et al. 2009). Extrapolating the findings from one individual to a whole population depends on how representative the marked individuals are. A single recovery or satellite track may not reveal what the population is doing.

2.2   Intrinsic Markers

The major advantage of an intrinsic marker is that it is not necessary to capture and mark an individual in advance, and every capture provides information on the origin of that animal. There is therefore, less bias than with extrinsic markers, where the location of the origin is dependent on accessibility; for example, remote areas might not be easily included in any migration study. Man-made contaminants such as dioxins and methyl mercury or other heavy metals could provide a means of inferring migratory origins as exposure to them varies throughout the world. Exposure of migratory animals to parasites or pathogens also varies geographically and could be used for determining movements but little research has been conducted as to how such markers might assist in deciphering migration patterns. In a study of Sharp-shinned Hawks (*Accipiter striatus*) in New Mexico, stable-hydrogen isotope ratios of feathers were estimated and blood was collected to quantify the prevalence and intensity of haematozoan infection in the birds (Smith et al. 2004). Twenty four percent of the birds were infected with *Leucocytozoon toddi*, 37 % with *Haemoproteus elani*, and 5 % with *H. janovyi*. This was the first documented occurrence of *H. janovyi* in North America in these hawks and the stable-hydrogen isotope analyses indicated that only birds originating from south-western North America harboured *H. janovyi* which could be of significance in regard to biogeography of that parasite species. The isotope studies also showed that birds infected with *Haemoproteus spp* were more widespread geographically than those with *Leucocytozoon* species. Another type of intrinsic markers that could potentially track animal movement geographically is the trace elements. Animals acquire quite distinctive chemical profiles relative to a particular geographical location and retain that profile when they move to another area. Trace element profiles have been used to determine if populations of breeding birds originate from different winter locations (Sze´p et al. 2008). However, there is little information on how trace elements vary over the globe, so it is of only limited potential. Stable isotope tools to trace animal migration have been used extensively in ecological studies but have not yet been utilized in work on the dissemination of transboundary animal diseases. There is potential for greater understanding the epidemiology of HPAI in relation to the migration of wild waterfowl by using stable isotope studies to identify the origins and stopover points of birds suspected of carrying HPAI. A summary of the advantages and disadvantages of extrinsic and intrinsic methods is given in Tables 1.1 and 2.1.

Table 2.1 Extrinsic markers for tracking animal migration. Reproduced with Permission from Hobson and Norris (2008)

Technique	Advantages	Disadvantages
Extrinsic	1. Can apply to a broad range of animals	1. Requires capture/recapture
	2. High spatial resolution	2. Biased towards initial captured population
Phenotypic variation	1. Inexpensive	1. Low resolution
	2. Can be used for historical specimen	2. Not applicable to all species
Banding/Marking	1. Inexpensive	1. Low recovery rates <0.5 %
	2. Gives information on start and end of migration	2. Data must be acquired over many years
		3. Only few banding stations across the world
		4. Marking and recovery are biased towards the banding station location
Radio-transmitter	1. Gives precise locations	1. Low range
	2. Gives precise trajectory	2. Expensive
		3. Transmitters might affect behaviour
Satellite transmitters	1. Precise animal trajectory	1. Expensive
		2. Only for use on large animals
		3. Transmitters might affect behaviour
International Cooperation for Animal Research Using Space ICARUS (www.IcarusInitiative.org)	1. Transmitter allows many different species to be tracked	1. High start-up investment
	2. Can track individuals over the globe	2. Technology not proven
Satellite tracking to determine the spread the spread of infectious diseases	3. Inexpensive after start-up costs	3. Transmitters might affect behaviour
Passive radar	1. Coverage over large geographical region	1. Coverage only from existing stations or portable instruments
	2. Inexpensive	2. Poor ability to determine species and individual movements
	3. Individuals need not be captured	
Transponders	1. Small size	1. Requires external activation
		2. Low range
		3. Coverage only form existing stations or portable instruments

(continued)

Table 2.1 (continued)

Technique	Advantages	Disadvantages
Geolocation tags	1. Produces animal trajectory	1. Individuals must be captured to download data
	2. Low weight	2. Accuracy relative to satellite tags low
	3. Inexpensive relative to satellite tags	

Studies of avian migration and population structure take advantage of geographical variation in stable isotope ratios, in particular the ratio of the two stable isotopes of hydrogen in precipitation. In North America, for instance, the ratio of growing season precipitation decreases from −20‰ VSMOW (Vienna Standard Mean Ocean Water) in Florida to −140‰ VSMOW in north-western Canada (McKechnie 2004). Because the deuterium of precipitation is incorporated into avian tissues via food webs and feathers are metabolically inert once grown, the deuterium values of feathers reflect the latitude at which they were grown. In species that moult prior to migration, feather deuterium can be used to identify the latitudinal origin of individuals, and hence link breeding and wintering grounds. Isotopic data can also be combined with information on capture dates or ring recovery. Stable isotope studies require a detailed knowledge of a species' moult schedule, and the assumption that the isotope signature of feathers reflects that of food webs in the region where the feather is grown must be verified. The technique has the best potential where there is a good spatial variation in the environmental locations in which the particular bird species is found.

The basic requirements in the use of stable isotopes are that:

- The characteristics of the environment through which the animal of interest moves are known. The term used to describe the mapping of large-scale and spatiotemporal distributions of stable isotope ratios in the environment and animals is an isoscape or isotopic landscape (Bowen and West 2008).
- Isotopic values in animal tissues can be discriminated from baseline isoscape values,
- The time period for acquisition into a particular animal tissue is understood. For instance, turnover of isotope will be greater in metabolically active tissues than in tissues with a slow, or no, turnover e.g. feathers (Table 2.2).

Just a handful of these stable isotopes are of any practical interest to studies on animal migration. These are the "light isotopes" of H, O, C, N, and S that are present in all components of the biosphere (plants and animals), the hydrosphere (water) and the atmosphere (gaseous O_2, N_2 and H_2O). These five elements comprise the bulk of all animal tissues (Table 2.3). These elements and their isotopes circulate in the biosphere to produce characteristic isotope distributions globally. Large pools of these elements provide stability in the overall isotope circulation e.g. in the ocean, for hydrogen and oxygen isotopes, inorganic carbon pool in the ocean for carbon

Table 2.2 Intrinsic markers for tracking animal migration stable isotopes. Reproduced with Permission from Hobson and Norris (2008)

Technique	Advantages	Disadvantages
Intrinsic	1. Not biased to initial capture population	1. Biased to final capture population
	2. Less labour intensive than most extrinsic methods	2. Lower resolution than extrinsic methods
Contaminants	1. Potentially high spatial resolution	1. Lack of distribution maps
		2. Transport of contaminants could give unreliable geographical signal
Parasites/Genetics	1. Several possible markers	1. Species specific
		2. Low resolution
Trace elements	1. Measurement of a large number of elements	1. Lack of distribution maps
	2. Potentially high spatial resolution	2. Expensive
		3. Requires more sample tissue
		4. Requires tissue that is metabolically inactive after growth
		5. Spatial resolution could be too high
		6. Some elements may be integrated into inactive tissue after growth is complete
Stable isotopes	1. Inexpensive	1. Low resolution
	2. Not species- or taxon-specific	2. No base maps
	3. Several isotopes can be combined to increase spatial resolution	3. Ideal tissue is metabolically inactive after growth
		4. Turnover rate of elements in active tissue is unknown
		5. Interpretation may be complicated by animals' physiology

isotopes, and sulphate in the sea for sulphur isotopes and the atmospheric reservoir for nitrogen. The presence of the stable isotopes varies widely in nature. All of the light isotopes have a common or abundant form, e.g. ^{1}H; 99.985 % and a more rare "heavier form", ^{2}H; 0.015 % (Table 2.3). The abundance ratios of these isotopes vary because of physical and chemical processes they undergo in nature and it is these variations that enable the use of stable isotopes in tracing migration. As it is difficult to measure the precise concentrations of the isotopes in samples, measuring the relative differences in isotopic ratios between a sample and a reference by means of a mass spectrometer is the way in which data are acquired. Since gas source isotope ratio mass spectrometers (IRMS) are used to measure light isotopes it is not possible to measure isotope ratios directly from a tissue sample – feather, blood, muscle, claw, hair – instead, the sample must first be combusted to form a gaseous analyte in an elemental analyzer, a gas chromatograph or a laser and the resulting gas can then be used to measure the isotopic ratios relative to a calibrated reference gas of

Table 2.3 Dry weight % abundance of light stable isotope ratios of interest in determining migratory connectivity in tissues. Reproduced with Permission from Wassenaar (2008)

Element	Weight (%)	Isotope ratios	δ Range (0/00)	Mass required (mg)
Carbon	30–40	$^{13}C/^{12}C$	05 to −65	0.2–1.5
Oxygen	27–40	$^{18}O/^{19}O$	+10 to +30	0.2–00.5
Nitrogen	12–19	$^{15}N/^{14}N$	−2 to +25	0.5–1.5
Hydrogen	6–8	$^{2}H/^{1}H$	−250 to +90	0.1–0.4
Sulphur	5–20	$^{34}S/^{32}S$	−20 to +30	1–2

Table 2.4 Average terrestrial abundances of the stable isotopes of major elements of interest in ecological studies the carbon cycle

Element	Isotope	Abundance (%)
Hydrogen	^{1}H	99.985
	^{2}H	0.015
Carbon	^{12}C	98.89
	^{13}C	1.11
Nitrogen	^{14}N	99.63
	^{15}N	0.37
Oxygen	^{16}O	99.759
	^{17}O	0.037
	^{18}O	0.204
Sulphur	^{32}S	95.00
	^{33}S	0.76
	^{34}S	4.22
	^{36}S	0.014

the same type. Development of IRMS equipment has led to automatic sample processing to obtain multiple isotope assays from a single sample, on decreasing sample size and improving throughput rates. IRMS can measure isotopic ratio differences to the sixth decimal place or ±0.01 % (Table 2.4).

The carbon cycle involves active exchanges of CO_2 among the atmosphere, terrestrial ecosystems and the surface ocean. The ^{13}C value of atmospheric CO_2 is decreasing in response to inputs of ^{13}C depleted CO_2 from fossil fuel plus biomass burning and decomposition. Over the past 100 years the decrease may have been almost 1‰, from about −7‰ to −8‰.

Depending on how plants' photosynthesis process is materialized, they are classified in two large groups, C3 and C4, with very different values for $δ^{13}C$. In the C3 group, the first photosynthesized organic compound has 3 atoms of carbon while in group C4, there are 4. Most plants (85 %) (E.g. trees and crops) follow the C3 photosynthesis pathway and have lower values of $δ^{13}C$, between −22‰ and −30‰. The remaining 15 % of the plants are of type C4. The majority are tropical herbs and have high values of $δ^{13}C$, between −10‰ and −14‰ (http://homepage.mac.com/uriarte/carbon13.html).

Carbon uptake by the dominant C3 plants on land involves a net fractionation of about 20‰ between the atmospheric CO_2 and plant biomass (−28‰). Carbon uptake by C4 plants, mainly tropical and salt grasses, involves a small net fractionation of about 5‰. Soil organic matter globally contains several-fold more carbon than either the atmosphere or living plant biomass and is similar or slightly enriched in ^{13}C in comparison with the dominant vegetation. The exchange of CO_2 between the atmosphere and the surface of the ocean involves an equilibrium chemical fractionation between atmospheric CO_2 (−8‰) and the total CO_2 in surface ocean water.

2.2.1   The Nitrogen Cycle

Most nitrogen in the biosphere is present as N_2 gas in the atmosphere. This massive reservoir is well mixed with an isotope composition that is essentially constant at 0‰. Nitrogen in most other parts of the biosphere also has an isotope composition near the 0‰ value, from −10 to +10‰, primarily because the rate of nitrogen supply often limits reactions such as plant growth and bacterial mineralization. Under these conditions all available nitrogen can be consumed, without regard to isotope content and with no overall isotope fractionation. Thus, slow rates of N supply and limiting amounts of substrate N are often important for understanding nitrogen isotope distributions. Some cumulative and large fractionations do occur in the nitrogen cycle. Lakes appear more variable in isotope composition than the large world ocean. Large isotope contrasts might be expected between lakes in which primary production is limited by N (little fractionation by phytoplankton) versus P (abundant N → large possible fractionations during N uptake by phytoplankton). Where phytoplankton have different ^{15}N values than terrestrial vegetation, the nitrogen isotopes may function as source markers for autochthonous and allochthonous organic matter.

2.2.2   The Sulphur Cycle

Sulphate in the ocean is a large well-mixed sulphur reservoir whose isotope composition is 21‰ heavier than primordial sulphur in the earth and solar system at large. Fixation of sulphate by phytoplankton occurs with a small isotope effect, but sulphate reduction in marine sediment occurs with a large effect of 30–70‰. Over geological time, and partially in response to global-scale fluctuations in sulphate reduction activities, the ^{34}S values of oceanic sulphate have varied from about +10 to +33‰. Uplift and preservation of marine sedimentary sulphides and sulphate-containing evaporates on land have produced a patchwork of sulphur in terrestrial environments, each with different ^{34}S values for bedrock sulphur. Thus, large ^{34}S ranges must be assigned in general sulphur cycle diagrams. In spite of this, continental vegetation seems to average near +2 to +6‰ over large areas and is quite distinct from the +~17 to +21‰ values of marine plankton and seaweeds. The stable isotope composition of sulphur entering the atmosphere can also be quite variable.

2.2.3   The Oxygen Cycle

There are three oxygen isotopes that act as tracers when the many common oxygen-containing molecules circulate in the biosphere. The water cycle controls much of the oxygen dynamics and oxygen isotope dynamics. Evaporation and condensation result in predictable variations in isotope compositions of water that are now routinely tracked at regional and global levels (http://isohis.iaea.org/userupdate/water-loo/index.html, http://www.waterisotopes.org/ and http://ecophys.biology.utah.edu/labfolks/gbowen/pages/Isomaps.html#IAEA).

Oxygen isotope studies with animals have focused on determining which local sources of water are used. The degree to which food influences ^{18}O variations in animals has not been determined fully.

2.2.4   The Hydrogen Cycle

Much of the hydrogen cycle involves water, with various processes in the water cycle leading to characteristic, large-scale geographic patterns of hydrogen isotopes in water. Ocean water is the main reservoir of hydrogen in the biosphere and the standard reference material "standard mean ocean water" for hydrogen isotope measurement. The isotope composition of ocean water is a good starting point for following isotope dynamics in the hydrological cycle. The transitions between liquid water and water vapour during evaporation and condensation involve kinetic and equilibrium reactions with isotope fractionation. Water vapour evaporating from the sea has ^{2}H values of -10 to $-20‰$, and as this process reverses during condensation and formation of rain and snow, this trend towards lower atmospheric ^{2}H values is amplified. As water vapour moves inland and up mountains, it progressively loses moisture and ^{2}H values decline further. These processes can be amplified yet again in colder regions where low temperatures promote stronger fractionations between vapour and condensate. A combination of high elevation and low temperature can result in ^{2}H values of -200 to $-400‰$ for water in high-elevation glaciers and for snow in Polar Regions. In less dramatic examples, large rivers fed by snowmelt that have continental origins can have much lower ^{2}H values than coastal marine waters. This makes ^{2}H source signals valuable tracers in coastal estuaries and floodplains linked to these rivers. Isotope hydrology studies often consider the water isotopes (hydrogen and oxygen isotopes) as markers for water sources and water circulation. Global-scale maps and animations of hydrogen and oxygen isotope variations in water are available on the Web (http://isohis.iaea.org/userupdate/waterloo/index.html or www.waterisotopes.org).

Analytical advances are making it easier to investigate the origins and cycling of hydrogen bound in organic matter. About 10–20 % of hydrogen in organic materials is exchangeable with water vapour present in normal laboratory air, but this exchange effect is understood and can be corrected for during routine analysis.

Hydrogen in animal tissues can be divided into three main pools: hydrogen derived from dietary sources, hydrogen from drinking water, and exchangeable hydrogen. For example, studies of quail fed deuterium-enriched showed that 10 % of all hydrogen was exchangeable, 30 % came from ingested water, and the remaining 60 % came from food. The overall finding for animals is that ^{2}H values are primarily controlled by diet, which in turn is strongly correlated with ^{2}H values of local water.

The reasons why hydrogen isotope tracers work well for studies of long distance migration depends on the fundamental chemistry of isotope fractionation. The water cycle of evaporation from oceans and precipitation inland involves isotope fractionations that leave behind the light isotopes. During chemical equilibrium fractionation, the vapour phase is enriched in the light isotopes, leaving the liquid phase heavier by difference or mass balance. The isotopically heavier liquid phase deposits as rain or snow, so that residual cloud-borne water moving inland or upwards is isotopically lighter and has lower ^{2}H values. Larger fractionations accompanying cold conditions in Polar Regions magnify some of these patterns, creating low ^{2}H values nearer the poles. This leads to continental-level patterns in the isotope compositions of water, in effect a giant isotope map created by the water cycle. Bird migrations take place across this chemical landscape, with birds at high latitudes having low ^{2}H values and birds near the equator having high ^{2}H values. The isotopes in the water provide a basic signal that first labels plants during photosynthesis and carbohydrate metabolism, then leads to general labelling of the local food web. The end result is that organic matter in materials such as bird feathers will have ^{2}H values that reflect the local water ^{2}H values. Migrating birds typically moult and form new feathers at the end of the summer, and feathers retain that late-summer isotope chemistry until the next year's moult.

2.3   The Stable Isotopes of Water on a Spatial Scale

As surface ocean waters evaporate into the atmosphere, the clouds formed are isotopically depleted in ^{2}H and ^{18}O relative to the ocean, resulting in an air mass that is similarly depleted relative to the ocean. In turn, as moisture is condensed from clouds during precipitation events, that water is isotopically enriched in ^{2}H and ^{18}O relative to the cloud, leaving the residual cloud mass isotopically depleted in ^{2}H and ^{18}O relative to the original cloud mass. The process of differential isotope depletion during precipitation results in a predictable pattern of depleted isotope ratios of precipitation as cloud masses move inland. Since the hydrogen and oxygen in precipitation become the primary source of H and O atoms incorporated into carbohydrates, proteins and lipids during microbe, plant and animal growth, these isotopes have the potential to carry a geographically based piece of information that are useful in migration studies.

The International Atomic Energy Agency (IAEA) and the World Meteorological Organization (WMO) has been conducting a worldwide survey of oxygen and hydrogen isotope content in precipitation. The objective is to collect basic data on isotope content of precipitation on a global scale to determine temporal and spatial variations of environmental isotopes in precipitation to provide isotope data for the use of environmental isotopes in hydrological investigations. To this primary objective two other objectives have been added, providing input data to verify and further improve atmospheric circulation models, and to study of climate change. Since 1961, more than 800 meteorological stations in 101 countries have been collecting monthly precipitation samples for the Global Network of Isotopes in Precipitation – GNIP, IAEA 2006. Precipitation samples are collected in cooperation with WMO, national meteorological services and national authorities. The samples are analyzed in the IAEA's Isotope Hydrology Laboratory in Vienna, and in cooperating laboratories. Each contributing laboratory is responsible for the accuracy and precision of its own analyses. The GNIP homepage can be accessed at: – http://www-naweb. iaea.org/napc/ih/index.html. GNIP data may be used freely, provided the source is cited as follows: – IAEA/WMO (2006), Global Network of Isotopes in Precipitation, The GNIP Database, (http://www.iaea.org/water). From these and other observations, both temporal and spatial patterns emerged that established the basis for a global representation of the distribution of isotopes in water on a global basis. Analyses of the spatial distributions of hydrogen isotopes of waters across North America and Europe reveal substantial variations in stable isotope ratios, making it possible to distinguish precipitation in many geographical locations on the basis of their ^2H and ^{18}O values. There are no unique stable isotope ratio values for waters in a specific geographic location, but rather gradients or bands of different isotope ratio values allowing one to distinguish between locations if they were sufficiently far apart from each other. Bowen and colleagues extrapolated from this location-specific data of stable isotopes in water to construct spatial maps of the predicted isotopic composition of water throughout the world so that it is possible to estimate reliably the globally averaged ^2H and ^{18}O values of precipitation for different latitudes and longitudes using a calculator available at http://waterisotopes.org.

There are a number of programs available online that can be used to estimate the mean annual and monthly H and O isotope composition of precipitation in different locations. They are intended to facilitate the use of stable isotope data, amongst other uses, in ecological studies, and to standardize the interpolation of precipitation stable isotope data. The program for calculating isotopes – The Online Isotopes in Precipitation Calculator, OPIC – is based on the work of Bowen and Wilkinson (2002), Bowen and Revenaugh (2003) and Bowen et al. (2005). These papers provide a description of the methods. The data that were used to develop the database used by the OIPC are derived primarily from the International Atomic Energy Association/(WMO) Global Network for Isotopes in Precipitation (GNIP). To use the database it is only necessary to input basic information on location and elevation to acquire ^2H and ^{18}O values for the specified location. The program is periodically updated as new data, including that provided from sources other than the Global Network for Isotopes in Precipitation, are added to the database. The OPIC can be accessed from the website managed by Purdue University, West Lafayette, Indiana, USA, (http://wateriso.eas.purdue.edu/waterisotopes/pages/data_access/oipc.html). When the database is used in publications, reference should be made to the database, The Online Isotopes in Precipitation Calculator, version XX (http://www.waterisotopes.org) and to publications by Bowen and Revenaugh (2003) and Bowen et al. (2005).

Another program that allows the user to download their own isoscape data is isoMAP, also based at Purdue University, that allows spatial analysis, modeling and prediction of stable isotope ratio variation in the natural environment It comprises web-based GIS and software tools that enable users to develop, and implement models for isotope distributions. This program can be accessed from http://isomap.rcac.purdue.edu:8080/gridsphere/gridsphere.

2.4   Deriving Isoscapes in the Absence of GNIP Data

Numerous studies have shown that precipitation isoscapes drive δD and $\delta^{18}O$ patterns in surface waters and in terrestrial food webs. While the GNIP dataset has provided the fundamental spatiotemporal foundation at a global geospatial scale, the GNIP stations are often spatially deficient for many regions of the planet that are of interest to ecologists. In North America, for example, Mexico has only two GNIP stations, which results in a spatial deficiency of water isotope data for that country. This spatial deficiency is of concern to researchers interested in large scale Mexican isotope hydrology, and to scientists interested in tracking migratory species to and from Mexico by using stable isotopes. Overcoming this GNIP spatial limitation would require adding spatially dense, long-term precipitation isotope collection stations across Mexico in order to obtain sufficient isotopic coverage of this topographically diverse country. This option is not viable for logistical and cost reasons, and so other approaches are needed to fill crucial information gaps. A possible approach is the application of the generalized globally predictive models for precipitation isoscapes (www.waterisotopes.org). However, these global models also lack sufficient or detailed on-the ground validation to determine how close they match or digress from reality at regional or country-wide scales (Bowen and West 2008).

In order to overcome this problem Hobson et al. (2009) developed a predictive general linear model (GLM) for hydrogen and oxygen isotopic spatial patterns in Mexican groundwater and then compared the results to a validation subset of field data, as well external data reported in the literature. The GLM used elevation, latitude, drainage basin (Atlantic vs. Pacific), and rainfall as the most relevant predictive variables. The GLM explained 81 % of the overall isotopic variance observed in groundwater, 68 % of the variance within this validation subset, and 77 % of the variance in the external data set. This predictive GLM is sufficiently accurate to allow for future ecological, hydrological and forensic isoscape applications in Mexico, and may be an approach that is applicable to other countries and regions where GNIP stations are lacking. They hypothesized that the stable isotopic composition of shallow phreatic groundwater in Mexico might serve as a useful proxy for integrating longer term (e.g. 5–10 year) precipitation infiltration inputs (Clark and

Fritz 1997). Groundwater infiltration is well-known to be a multi-year integrator of seasonally weighted rainfall events (Clark and Fritz 1997; Darling and Bath 1988; Rozanski 1985), and hence its isotopic composition closely reflects that of seasonally weighted long-term average inputs even in some of the most arid regions of the world (IAEA 2007). Groundwater was sampled from 234 sites at ~50 km latitudinal spacing to obtain high spatial resolution and country-wide coverage for the construction of a groundwater isoscape. Shallow groundwater infiltration in Mexico appeared largely unaffected by evaporation and reflected seasonally weighted precipitation inputs.

2.5   Use of Stable Isotopes for Migration Studies

The stable isotopes of H, O, N, C and S are routinely studied in bird migration because they are recorders of the dietary sources of the birds and these sources can be interpolated or specifically linked to ground proven or large-scale patterns in the global landscape and water environments (Hobson and Wassenaar 1997). Depending on whether the tissue of the migrating animal is biochemically fixed (feather, hair) or dynamic (blood, muscle) these stable isotopes provide fundamental information about where the animal has been and what it has been eating. The technology is relatively recent in inception and although it has been widely used approaches to its use are evolving, techniques are improving and methods of interpretation are being updated constantly. Stable isotopes as intrinsic markers do not require a recapture of the same animal to obtain data since the required spatial information is retained in the animals' tissues as a result of its stay in a particular habitat. Stable isotopes depend on several key requisites in order to function as intrinsic markers. Firstly, the animal must acquire a stable isotope into its tissues; since the light stable isotopes are key building blocks in animal tissues this is readily achieved. Secondly, the migrating animal must move between different environments and retain in its tissues measureable isotopic differences that can be linked to diet at previous or current locations. This is easily met by species that migrate seasonally across distinct isotopic landscapes such as between higher northern latitudes and lower latitudes in the southern hemisphere. It also requires that isotopic discrimination between the different seasonal landscapes and the tissues being measured is consistent or well-known, or has been empirically measured. Such conditions may not always be met with and provide considerable challenge to the interpretation of results. Any migration study related to HPAI will need to be conducted collaboratively between virologists, biologists and ecologists with experience in bird migration and scientists with access to a stable isotope laboratory facility. Stable isotope analysis is a costly procedure and it is essential that appropriate links are made that will ensure the right approaches to determine migratory connectivity are made and that a well-designed project that will answer the questions on virus dissemination and wild bird migration.

The IRMS used most frequently for SIA is the continuous flow isotope ratio mass spectrometer (CF-IRMS). These machines enable relatively low cost analysis, a high degree of automation and a high throughput. A technique known as comparative equilibration using CF-IRMS is used for δD analysis for migration research. This approach allows comparative results between different laboratories, ease of use and rapid, automated sample throughput of samples (Wassenaar and Hobson 2003). It requires the inclusion of pre-calibrated keratin working standards along with the unknown tissue samples. The hydrogen isotope exchange between ambient moisture in the air and the keratin standard and the samples is identical. The comparatively equilibrated samples and standards are then isolated from the atmosphere and analyzed together in a single session. Measurement of stable hydrogen in tissues is performed on H_2 derived from high-temperature flash pyrolysis. Pure H_2 is used as the sample analysis gas and the isotopic reference gas. A high temperature Elemental Analyzer and autosampler pyrolyzes samples to H_2 gas. The pyrolysis column consists of a ceramic tube filled with glassy carbon chips held at 1250–1350 °C, followed by molecular sieve gas chromatography column at 80–100 °C. The IRMS makes measurements in the following way: gases enter a white-hot region where electrons are boiled off and the sample gases are ionized. These ionized molecules have a positive charge and they are passed through a magnetic field that separates them according to their atomic mass and isotopes with the resulting ion beams focused into collectors for counting. Computers then calculate the final isotope values. The stable isotope ratios are then reported as a series of delta (δ) notations of positive or negative numbers expressed as parts per thousand (‰) relative differences to an international standard. All δD results are expressed in units per mil (‰) relative to the Vienna Standard Mean Ocean Water using the previously calibrated keratin standard.

One of the most important practical issues in using δD measurements to track large-scale terrestrial animal movement is the problem of hydrogen exchange in organic samples It is well established that δD measurements of feathers (and other organic materials) are problematical compared to $\delta^{13}C$ and $\delta^{15}N$ analyses due to the problem of uncontrolled hydrogen isotopic exchange between "exchangeable" organic hydrogen in the feathers and isotopically variable ambient moisture in the laboratory environment. If left uncorrected, δD measurements of the total hydrogen in an identical feather will yield different results at analytical laboratories located at different geographic locations, as well as over time due to geographical and seasonal changes in the hydrogen isotopic composition of ambient moisture. This would make it impossible to compare δD results between different laboratories. Wassenaar and Hobson (2003) showed how keratin δD analyses by CF-IRMS technology and the use of keratin standards and comparative equilibration can be used to provide cost effective and rapid δD analyses for migration research to provide cost effective and comparable δD results on feathers and other keratin-based samples among different laboratories. Three in-house keratin standards were prepared from chicken feather cow hoof and bowhead whale baleen. Each keratin type came from a single geographic location. All of them were solvent cleaned, dried, cryogenically ground, and homogenized. They were then analyzed 6 times each using the offline equilibration dual-inlet method described by Wassenaar and Hobson (2000).

The target weight of feather (and in-house keratin standard) required for routine online hydrogen isotope analysis by CF-IRMS in our laboratory was 350 ± 10 µg. Feather samples were cut from the same position on each feather and weighed using a microbalance, and transferred into 4.0 mm 3.2 mm isotope grade silver capsules. Capsules containing the keratin standard powder or feather were folded into tiny balls. All weights were recorded and the samples were stored in 96-well ELISA plates, loosely covered with the lid. Samples and keratin standards were then allowed to "air equilibrate" on the shelf with ambient lab air moisture at room temperature for >96 h prior to stable hydrogen isotope analysis. The hydrogen exchange process could be sped up by holding the samples and working standards in an oven at higher temperatures since showed that complete exchange with moisture occurs in less than 30 min at temperatures above 100 °C. For practical purposes, however, a 96-h equilibration at room temperature was deemed sufficient and easier, and could be easily implemented as a standard laboratory analysis procedure. Following 96-h comparative equilibration, all samples and standards were immediately loaded into the auto sampler carousel of the CF-IRMS and analyzed for δD.

The uncorrected hydrogen isotope values of the keratin standards from each CF-IRMS auto run were subjected to a least squares regression to derive a correction formula to be applied to all the keratin standards and feather samples within that auto run. A typical run included an in-house keratin standard at the beginning, 2 keratin standards after every 10–12 unknowns, and 2 or 3 keratin standards at the end of the auto run. The authors recommend that researchers performing δD analyses adopt this protocol in order to provide findings that are comparable among laboratories. They also suggest that the development and distribution of feather and keratin working standards that have a wide range of δD can be done on an individual basis, or preferably, developed and distributed as a future collaborative effort among avian researchers and stable isotope laboratories that provide feather δD analyses. Since CF-IRMS enables δD analyses to be conducted on small amounts of tissue samples it is possible intra-sample variance in the hydrogen isotopic due to any biological heterogeneities could exceed interpretations of environmental isotope. To help resolve this, feathers were obtained from captive birds to examine isotopic variance expected due to sample size, location, and heterogeneity factors, and from selected wild birds to examine isotopic variance due to these and to additional dietary or location changes during feather growth (Wassenaar and Hobson 2006). Captive bird feathers were sub-sampled along the vane on either side of a single feather at masses of 0.25, 0.35, 0.45, 0.6, 1.0 and 2.0 mg, and along the rachis. The results showed consistency of feather δD measurements across a wide range of sample masses. Within-feather δD isotopic variance for captive and some wild birds was as low as ±3‰ for vane material, which corresponds to a geospatial resolution of about 1° of latitude in central North America. Intra-sample variance for the rachis was ±5‰, with lower δD values for both wild and captive birds. Nevertheless, the authors recommend researchers assess the degree of intra- and inter-sample hydrogen isotopic variation in the selected tissue growth period for the species of interest before geospatial interpretations of origin are attempted.

2.6   Approaches for Determining Migratory Connectivity

Patterns in stable isotopes derived from the environment are translated into food webs and the discrimination between stable isotopes and dietary sources and the feather is predictable and constant in time and space. The isotopic values of an animal are assumed to be representative of the location at which the feathers was grown and when the bird moves elsewhere it can be sampled to infer its previous geographic location. In order to infer the location most likely associated with the feather isotopic values it is necessary to calibrate the geographic model using tissues from birds of known origin. The accuracy of the IRMS measurements is best when using standards with chemical composition similar to the unknown samples (e.g. keratin standards to calibrate feather samples). Also, for geographical assignment it is best that the standards have similar attributes, i.e. the same species and isotopic values as the unknown samples. A calibration dataset should include samples from all areas from which the migrant of interest could have originated. However, sampling tissues from all potential places of origin is costly and logistically difficult if not impossible if the areas are remote and beyond the reach of sampling and collection networks.

The simplest method of assigning a bird to an isoscape is to define the geographic gradients based on maps of isotopic values, measure an individual and then assign it to the area that corresponds to the isotope values obtained. The spatial patterns are generated from indirect sources of information such as rainfall. The earliest studies on migration utilized this technique (Hobson and Wassenaar 1997) to use δD to study migratory birds and it is still in common use (Hobson et al. 2006, 2007). The method is simple to understand and apply and no additional computation is required to place birds in their geographical locations. More advanced methods for determining the origin of migratory animals require some degree of statistical manipulation. Assignment methods require the definition of all possible locations of origin, then characterizing these locations with the distribution of stable isotope data. The characterizing should, if possible, be derived from measurements of birds known to have grown tissues at those locations and all possible regions of origin are sampled. Once a calibrated assignment model is obtained, stable isotope measurements from individuals of unknown origin are used to determine their origins.

Although migratory individuals have only to be captured once in order to determine their migratory movements the assignment of their origin is challenging and it is necessary to make inferences about the history of an individual using the stable isotope measurement. Table 2.5 details some of the different modelling approaches that have been applied to estimate origin when using stable isotopes.

Table 2.5 Methods for assigning animals of unknown origin using stable isotopes. Reproduced with Permission from Wunder and Norris (2008a)

Level of complexity	Model type	Description	Advantages	Disadvantages	Incorporates sources of error	References
Low	Map lookup	Isotope value of animal assigned to area based on isoscapes on map	Easy to implement	Does not incorporate spatial variability of isotopes	No	Hobson et al. (2007)
	Linear regression	Origin inferred on regression of isotopes on latitude and/or longitude	Easy to implement	Does not incorporate spatial variability of isotopes within regions	No	Kelly et al. (2002)
	Classification trees	Origin inferred on basis of hierarchical, discrimination based decision rules	Can be applied to multiple isotopes. Does not require distributional assumptions	Does not incorporate most error, no a priori hierarchy for multiple isotopes does not provide degree of certainty for branching decisions	No	Hebert and Wassenaar (2005)
	Likelihood-based assignment	Origin inferred from probability density functions for isotope values from given regions	Can be applied to multiple isotopes. Provides probability of assignment for a given individual, easy to implement	Does not incorporate most error, simply assigns regions for individual based on highest likelihood value; requires sampling all potential regions of origin	Some	Kelly et al. (2005)

Level of complexity	Model type	Description	Advantages	Disadvantages	Incorporates sources of error	References
	Likelihood with priors	Same as above but adds prior information on isoscapes using Bayes' Rule	Can be applied to multiple isotopes and can utilize non-isotopic information	Does not incorporate most error. Simply assigns region for individual based on highest posterior probability; requires sampling all potential regions of origin	Some	Norris et al. (2006)
	Stochastic extension of likelihood	Same as above but adds sources of error associated with isotope data	Can be applied to multiple isotopes, incorporates multiple sources of error and provides a range of assignments for a given individual	Requires intensive computing	Yes	Wunder and Norris (2008b)
High	Probability Surfaces	Models stochastic error process over mean field surface	As above, incorporates known variance sources	Requires intensive computing	Yes	Wunder (2010)

2.6.1 Classification Trees to Predict Origins

Ecological data are often complex, unbalanced, and contain missing values. Relationships between variables may be strongly nonlinear and involve high-order interactions. The commonly used exploratory and statistical modelling techniques often fail to find meaningful ecological patterns from such data. Classification trees are statistical techniques ideally suited for exploring and modelling such data. They have a number of advantages over discriminant function analysis and linear regression, both of which are often used in stable isotope assignment studies. Classification trees represent a statistical technique well suited for modelling ecological data since the model development is hierarchical and based on logical if-then conditions and are both nonparametric and nonlinear (De'ath and Fabricius 2000). Hebert and Wassenaar (2005) used the classification tree approach to determine if ducks originating from different geographic areas had unambiguous multi-isotopic signatures. They used a multi-stable isotope analysis ($\delta^{34}S$, δ^2H, $\delta^{13}C$, $\delta^{15}N$) of secondary feathers from wild duck (*Anas platyrhynchos*) and northern pintail (*A. acuta*) ducklings from 52 sites in western North America. Ducklings from Alaska, northern Canada, the Prairies and California could be distinguished based upon their feather isotope values. Classification trees were developed by repeatedly splitting the data via algorithms that partition the data into mutually exclusive groups (De'ath and Fabricius 2000). Geographic patterns in feather isotopes were related to natural gradients produced by biogeochemical cycles and anthropogenic factors such as agrochemical usage.

2.6.2 Likelihood-Based Methods to Predict Origins

Discriminant analysis can be used to classify a sample into two or more classes (regions) and has been employed in determining the origin of birds. Parameters for region-specific likelihood functions are estimated from isotope data collected in each predefined region. The likelihood functions for each region are then evaluated for the isotope value measured for an individual of unknown origin and that individual is placed in the region with the highest-valued likelihood (or probability).

In studies on golden plovers (Prosser et al. 2006) interpolated maps were used of annual average δD values collected at stations across the wintering grounds of each of two species of plover to generate a priori predictions for changes in δD in feathers between breeding and wintering grounds (IAEA 2001; Bowen et al. 2005). Because there was no information on the specific wintering location of each bird, the expected range of δD feather values was evaluated from these maps (Bowen et al. 2005). The annual averages of deuterium in the precipitation across wintering areas were used to predict feather δD values. To assign individuals to group of origin, the principle of "assignment with exclusion," was used, a method adopted from population genetics that uses an exclusion criterion to reject unlikely sources. The cross-validation procedure in discriminant analyses as an assignment method was used to classify feathers into known group of origin based on the three isotopic values. The limited number of sample contributed to low power of some assignments, but power was increased by using three isotope values for each sample.

2.6.3   Migration Studies Using Stable Isotopes

The first premise for using stable isotopes is that there are differences in the spatial distribution of the isotopes from where the migrants obtained their diet. The stable isotopes δD and $\delta^{18}O$ have the most predictable geographic gradients, but gradients for other light isotopes are less well known. Geographic patterns of stable isotopes are inferred from a number of sampling points across the landscape. The δD values in precipitation have been modelled for North America (Hobson and Wassenaar 1997) where the relationship between δD in feathers and growing season precipitation was examined for neotropical migrant songbirds breeding over a continent-wide isotopic gradient. The δD values were determined on feathers of 140 individuals of 6 species of wild insectivorous forest songbirds (*Setophaga ruticilla, Empidonax minimus, Vermivora peregrinus, Catharus ustulatus, Seiurus aurocapillus, Hylocichla mustelina*) taken from 14 breeding locations across North America. The δD of feathers was strongly correlated with the δD of growing season precipitation at breeding sites across North America. As feather hydrogen is metabolically inert after growth, this relationship was then used to assess the breeding origins of wintering migrants. Deuterium values of feathers from 64 individuals representing 5 species of migrants (*Helmitheros vermivorus, Wilsonia citrina, Hylocichla mustelina, Dumetella carolinensis, Seirus aurocapillus*) at a wintering site in Guatemala were consistent with those predicted from the known breeding ranges of these species. Bowen et al. (2005) described a method for interpolation of precipitation isotope values to create global base maps of growing-season (GS) and mean annual (MA) δD and $\delta^{18}O$. The use of these maps for forensic application was demonstrated using previously published isotope data for bird feathers in North America and Europe. The precipitation maps show that the greatest potential for applying both hydrogen and oxygen isotope forensics exists in mid- to high-latitude continental regions, where strong spatial isotope gradients exist. They showed that feather δD/precipitation δD relationships have significant predictive power both in North America and Europe, and zones of confidence for the assignment of origin could be described using these predictive relationships. These maps are available in GIS format at http://www.waterisotopes.org.

2.6.4   Determining Migratory Connectivity for Waterfowl in Asia

Although Asia is a prime focus for HPAI, and despite considerable interest in conservation and disease surveillance issues involving waterbirds in that region investigations on migratory connectivity in wild birds that might be implicated in the spread of the disease are few (Chang et al. 2008; Pérez et al. 2010). The great cormorant (*Phalacrocorax carbo sinensis*) is a potential carrier of HPAI to Taiwan and to better understand the migration of this bird, SIA studies were conducted on several populations from different breeding sites and overwintering grounds (Chang et al. 2008). Rather than capturing and removing feathers from live adult birds, newly dropped feathers found lying on the ground including primary, secondary and tail feathers at collected from three breeding sites in China (Qinghai Lake), and Russia (Ussuri River, and Chita Peninsula) and overwintering grounds in Taiwan (Kinmen Lake). All the collected feathers were newly dropped because these feathers were neither weathered, nor damaged, nor covered by a layer of dirt or dust. They were collected from different individuals by collecting them under widely separated nests at breeding sites and under separated roosting trees. Although δD values of feathers have typically been compared with the δD of precipitation to infer migration, the distributions of δD and $\delta^{18}O$ in precipitation across Asia are not as precisely known as in North America and Europe as there are fewer data collection sites and the frequency of data collection is more sporadic in Asia. This makes it more difficult to infer the origins of birds by comparing the δD values of precipitation and of feathers in Asia alone hence, a direct comparison of $\delta^{13}C$, $\delta^{15}N$, $\delta^{18}O$ and δD values of feathers collected at both wintering and breeding sites was used to infer breeding locations of *P. c. sinensis* wintering at Kinmen. Analysis showed that it was unlikely that *P.c.sinensis* wintering at Kinmen came from the breeding sites in China and Russia.

Analysis of the data collected from the *P.c.sinensis* wintering at Kinmen, produced nine clusters plus two outliers and approximately 65 % of the birds overwintering at Kinmen come from four breeding populations. Using $\delta^{13}C$, δD, $\delta^{15}N$ or $\delta^{18}O$ values, excluded the possibility that cormorants wintering at Kinmen came from breeding populations at Qinghai Lake, Ussuri River or the Chita Peninsula. The general locations of possible breeding sites were based on the comparisons of feather δD values with the mean δD of precipitation during the breeding season according to the best available δD isocline maps of Asia, from which the data are spatially and temporally discontinuous. This comparison suggested that approximately 50 % of the *P.c.sinensis* wintering at Kinmen come from five breeding populations from the region around Lake Baikal or an area encompassing the Amur, Khabarovsk and Primorsky regions of Russia. The study showed that SIA has the potential to provide migratory information at the population level, to uncover the possible existence of previously unknown breeding populations, and to make predict migratory and possible future transmission pathways of HPAI. Moreover, combining multivariate analyses of stable isotopes with analyses of the phylogenic trees of birds and HPAI may provide an opportunity to better understanding the interaction of bird migrations and the spread of HPAI.

In another study (Pérez et al. 2010) feather sample were collected from Bar-headed Geese (*Anser indicus*), Whooper Swans (*Cygnus cygnus*), Mongolian Gulls (*Larus vegae mongolicus*), Curlew Sandpipers (*Calidris ferruginea*) and Pacific Golden Plover (*Pluvialis fulva*) at seven sites in northern Mongolia. After cleaning to remove oils, stable-hydrogen isotope (δD) analysis of one feather per individual was conducted at the stable isotope facility at the National Hydrology Research Centre, Saskatoon, Saskatchewan, Canada. Isotope analyses followed the comparative equilibration technique described by Wassenaar and Hobson (2003). Deuterium isotope ratios were expressed in δ notation in parts per thousand (‰) relative to the Vienna Standard Mean Ocean Water Standard Light Antarctic Precipitation (VSMOW-SLAP) scale. Measurement precision based on replicate measurements of within-run keratin standards was estimated to be of the order of ±2‰. Estimates of continental deuterium patterns in mean growing-season precipitation (δDp) for Asia were derived from Bowen 2009 (www.waterisotopes.org). Continental δDp data show an expected general decrease in deuterium abundance in precipitation with increasing latitude and altitude in Asia and form the basis for predicting the expected abundance of deuterium in feathers (δDf) from the average δDp (Hobson and Wassenaar 1997; Hobson 2008). Three major feather isotopic source areas in Asia were identified: Region A – northern Asia (<−138‰), B – catchment area (−138 to −104‰) and C – southern Asia (>−104‰) were delineated. The isotopic interval of 34‰ presented for Region B corresponded to latitudes ranging from approximately northern China to mid-Russia, and Regions A and C were north and south of those latitudes, respectively. These arbitrary cut-off values in δDf were chosen because they bracketed Region B between the lowest and highest 95 % CI out of the expected mean annual δDp values for the latitudes, longitudes and altitudes of the sampling sites. Once more precise information on the relationship between δDp and δDf can be established for Asia, more refined assignment will be possible (Hobson et al. 2009). Feathers from adult Pacific Golden Plover δDf had values of −33‰, clearly indicated that these feathers were moulted in southernmost Asian latitudes, suggesting that the feathers were probably grown on or near the wintering grounds prior to spring migration. Locally grown feathers from adult Barheaded Geese and Whooper Swans had more depleted δDf values than any other group because they live in a complex of lakes and ponds fed by high elevation run-off, and the waters are more depleted in deuterium than adjacent areas. Curlew Sandpiper had δDf values consistent with Region B indicating isotopic equilibrium with diet on their sampling area. The study demonstrated that stable isotopes could be used to improve knowledge of migratory and moulting patterns of wild waterbirds in Asia in a rapid and cost effective manner. However, additional studies are required since the δDp isoscape for Asia is so poorly described. Ground-truthing the relationship between δDf and δDp over a large geographic gradient in Asia will provide more information to derive appropriate isotopic calibrations (Wunder and Norris 2008b; Wunder 2010).

Arrival time on breeding or non-breeding areas is likely to be of interest in epidemiological studies exploring dissemination of HPAI by migratory birds. Precisely assessing the arrival time of individuals can be difficult, but by measuring carbon stable isotope turnover in avian blood it is possible to estimate arrival time for birds switching from one habitat to another (Oppel and Powell 2010). Stable carbon isotope ratios (δ^{13}C) in blood assimilate to a new equilibrium following a diet switch according to an exponential decay function. Stable carbon isotope ratios in tissues reflect the isotope ratios of food sources (Hobson and Clark 1992). When migratory birds switch habitat their new diet will have a different isotopic signature if the two habitats are isotopically distinct (Peterson and Fry 1987). This switch to a new diet causes the isotope ratios in blood to change gradually over time until they reflect the isotope ratio of the new diet (Hobson and Clark 1992; Evans Ogden et al. 2004; Morrison and Hobson 2004). The rate of change in δ^{13}C is tissue dependent. For example, blood plasma generally assimilates to the stable isotope ratio of a new diet within a few days (Hobson and Clark 1992), whereas whole blood or the cellular fraction of blood (red blood cells, RBC) turn over within several weeks (Bearhop et al. 2002; Evans Ogden et al. 2004; Morrison and Hobson 2004). Several experimental studies have determined that isotopic turnover in blood closely follows an exponential decay function (Evans Ogden et al. 2004; Carleton and Martinez del Rio 2005). This relationship can be used to determine the time an animal switched diets if the isotope ratios of the old and the new diet are known.

A study of the arrival time of Eider Ducks (*Somateria spectabilis*) at breeding grounds (Oppel and Powell 2010) was estimated using only a single tissue (Phillips and Eldridge 2006). Since the turnover rate of stable carbon is not known in many bird species a mass dependent turnover rate constant was utilized in calculating the arrival time (Carleton and Martinez del Rio 2005). Data from whole blood and RBC measured in captive experiments were used to validate the approach of estimating the time since a diet switch. In each experiment, birds were switched from one isotopically distinct diet to another differing by at least 3 % in δ^{13}C (Hobson and Clark 1992, 1993; Bearhop et al. 2002; Evans Ogden et al. 2004). This provided data to develop a formula that could be used to determine arrival time. If the isotope signatures of a tissue from both the old and new environment are known and birds are captured at an unknown time after arrival, then blood δ^{13}C can be used to determine the time since elapsed since arrival. The isotopic signature of each diet can be characterized by sampling blood from birds that have been feeding in either environment for several weeks, or by using tissues that are metabolically inert after growth. Feathers sampled from the bird may be useful to determine δ^{13}C of the previous environment if the moulting strategy of the bird is sufficiently known and certain feathers are always grown in the previous environment. Plasma or RBC sampled from a bird captured at a later stage can be used to determine new δ^{13}C once birds have reached equilibrium with the new diet (Morrison and Hobson 2004). For breeding grounds, eggshell membranes collected from old nests provides a simple way to characterize the diet of birds (Oppel and Powell 2010).

Stable isotopes may also provide clues to help locate nonbreeding populations of birds that have unknown winter ranges. For instance, no records exist for the Coastal Plain Swamp Sparrow (*Melospiza georgiana nigrescens*) from October to late April because of the difficulty in tracking individuals between seasons. Stable isotope analyses of C, N, and H were used to predict where moult occurs and then those areas were then searched for individuals of this species (Greenberg et al. 2007). Feathers were clipped from the crown and rump and lower back of the birds. Analysis of the crown feathers was used to predict the δD value at the unknown site of winter migration. Selection of the sites was based on the estimates of isotopes in precipitation reported by Bowen and colleagues (www.waterisotopes.org). The δ^{13}C and δ^{15}N of rump feathers were consistent with the Sparrow moulting in more saline marshes. The values for the same isotopes from crown feathers revealed that winter moult in the Sparrow probably occurred in similar coastal brackish habitats. The δD of crown feathers indicated that pre breeding molt occurred at latitudes between South Carolina and Virginia. A subsequent search of this region located specimens of the birds, all of which were found in North Carolina or southeastern Virginia. The Sparrows were found predominantly in brackish marshes similar to their breeding habitat. On the basis of these observations, it appears that Coastal Plain Swamp Sparrows undergo a short southerly migration to a coastal region with substantially warmer winter conditions. This study was the first to make a specific geographic prediction based on stable-isotope analysis and to test the prediction in the field.

Many birds that pass through Africa on their migration cross the Sahara or Arabian deserts without delay, then stopover in northern tropical areas for several months before resuming their flights to the south. During the stopover they complete a partial or complete moult (Yohannes et al. 2007). Stable nitrogen (δ^{15}N), carbon (δ^{13}C) and hydrogen (δD) isotope profiles in feathers of nine migratory bird species trapped in Kenya were examined to test the extent to which they were segregated, geographically or by habitat, during their earlier autumn migration stopover in northeast Africa. The aim was to determine if isotopic differences between species varied between years, and whether the isotope profiles of individual species appeared to be consistent. The relationship between mean feather δ^{13}C, δ^{15}N and δD assorted the migrants into several clustered groups. Similar feather isotope values among successive years revealed that each species tended to return to the same or similar stopover areas and selected habitat and diet that generated similar isotopic signatures. However, the stopover sites that the birds used were not identified.

Chapter 3
Practical Considerations

3.1 Sample Collection and Tissue Preparation

In assessing the value of stable isotopes for tracing animal migration the O and H isotopes in the environment are "global spatial" in their resolution as the isotope are related to hydrological and meteorological processes that are seasonally and spatially predictable over many years on a regional, continental and global scale. This enables interpolation into regions where no long term data are in existence (Bowen et al. 2005). Isotopes of C, N, and S are local spatial in character as they do not vary over the landscape as H and O. investigations into large scale migrations is therefore most fruitful using H and O isotopes, while the resolving power at the local habitat level can be improved with data on the local spatial isotopes.

Two types of samples that can be used for analysis; the first type is fixed tissue consisting of body material that once formed is metabolically and therefore by extension, isotopically inert. This material will give an indication of the isotopic composition pertaining in the local environment at the time that the tissue was formed. Tissues of this type include keratinous material of claws, nails, hair and feathers. Bird feathers are particularly well-suited for use in tracking migration as the fully formed feather vanes and rachis do not change chemically or isotopically as the bird moves away from the site of feather formation, on its migration pathway. The feather is often grown over a short period of time at a specific site such as a breeding site or wintering ground and so will retain the isotopes present in the diet at that location. It is essential that the researcher has adequate knowledge of the ecology and physiology of the birds being studied. For instance, slowly growing feathers or claws that grow as an animal migrates, ingesting different diets en-route means that a growing feather will record a variation in isotope as it progresses. In bald eagles where flight feathers grow as they migrate from the northern to the southern USA, there is intrasample variation where the feather show a negative δH value in the oldest part of the feather at the wing tip, while at the base of the feather the δH values are more positive. In this case the feather is not equivalent to a fixed

The original version of this chapter was revised: The chapter was inadvertently published with materials that were reproduced or modified without permission for the table headings in Table 3.1 which has been updated now. The correction to this chapter is available at https://doi.org/10.1007/978-3-319-28298-5_4

tissue and provides instead a dynamic spatial recorder. In the case of a migratory bird in which the feather is formed at the natal location there would be no significant intrasample differences in δH that would indicate where the bird had moved, even though captured far from the region in which the feather was formed. Archived or museum material is also a potential source of investigation for stable isotope analysis, but care must be taken in using such material if it has been preserved or contaminated in some way whilst in storage.

The second type of sample material is dynamic tissue, metabolically active tissues in which the stable isotope composition reflects the current, active dietary habitat. These tissues are the blood, muscle, liver etc. There have been few studies using such tissues to plot migratory connectivity. Most studies have concentrated on measuring $δ^{13}C$ and $δ^{15}N$. The turnover rate for isotopes in different tissues varies from a few days in liver and blood, to weeks in muscle to a lifetime in bone collagen.

An important consideration is whether the sample can be obtained non-lethally by plucking or cutting feathers, nails or hair so as not to inconvenience the animal. Animal care guidelines should always be exercised when sampling live animal in the field. Collecting a sample for analysis is a major undertaking and it is essential that due consideration has been given for the animal's biology and ecology to ensure that appropriate samples are collected.

3.1.1   Questions of Sample Heterogeneity

For birds, flight feathers or contour feathers can be sampled for isotopes, but to if replicate samples are obtained what extent would isotopic values vary from the isotopic range that defines the bird's spatial distribution? There will be individual and population level intersample isotopic heterogeneity. There could be minor variation in values between contour feathers grown by an individual at a single location. There would as also be intersample isotopic variation for the same feathers obtained from a population of birds that grew the feathers at the same location. This sort of heterogeneity is usually much less than the large scale geospatial isotopic patterns. Intrasample isotopic heterogeneity is the variance seen in a single discrete sample; an example of this is as mentioned above for the bald eagle.

3.1.2   Preparation of Feather Samples

Samples collected from the field are likely to be dirty and greasy. Dirt can be removed by washing in distilled water but surface oils on feathers will need to be removed by organic solvents since the composition of H in the oils differs from that contained in the feather keratin, thereby giving false data on isotope composition.

Table 3.1 Procedure for preparing feather samples for δD stable isotope analysis. Reproduced with Permission from Wassenaar (2008)

Procedure
1. In order to remove surface oils clean feathers in 2:1 v/v chloroform/methanol by soaking for 24 h and rinsing twice in same mixture. Air-dry in a fume hood for at least 48 h
2. Cut off a small amount of feather vein, not the central rachis, for analysis. Always cut samples from same region on different feather samples to ensure consistency (e.g. sample at tip). Cut feathers using stainless steel scissors and forceps
3. Clean instruments using methanol and tissue wipes and allow to dry. Do not use acetone
4. Clean and calibrate microbalance
5. Tare a silver capsule* handling it only with instruments, remove and set on a metal surface. Use the smallest size that will safely hold a sample (3.5×5.0 mm)
6. Transfer a small amount of feather to capsule using forceps. Reweigh and add or subtract material until the target sample weight of 350±10 μg is achieved. Reference materials must be weighed to a comparable elemental mass as the samples
7. Seal the capsule by crimping the end shut with straight edge forceps and then fold down tightly as if rolling down a bag. Use forceps to gently compact the capsule into a small tight cube or ball. There should be no loose sides, stray edges or feathers poking out. Flattened samples can jam the autosampler
8. Record sample weight and sample name in an Excel spreadsheet conformed in a 96-well ELISA template format. Put capsule in appropriate position in ELISA plate. Clean all instruments with methanol and tissue wipes after completing each sample. Put lid on ELISA plate and secure with rubber bands or tape to ensure samples do not jump out of wells in later handling. Record sample name etc. and weight and its position in the ELISA plate. Transfer information to an isotope laboratory submission form

*Silver Capsules must be used for analysis of $\delta^{18}O$ and δD. They can be obtained from: (1) Silver capsules for solid samples, 3.5×5 mm, 041066 240-054-00; Costech Analytical Technologies Inc.; 26074 Avenue Hall, Suite 14; Valencia, CA 91355; USA or (2) D2003 – Silver Capsules Pressed 3.5×5 mm; OEM Parts: E12050, 041066, SC1082; Elemental Microanalysis; 1 Hameldown Road; Okehampton, EX20 1UB; UK

Samples collected for stable isotopic analysis must be stored properly both before and after preparation to ensure there is no degradation. Many laboratories can provide a service but if the researcher wishes to reduce costs then cleaning and weighing can be undertaken. This is not to be undertaken lightly as accurate weighing to ±10 μg is required using analytical microbalances. A recommended procedure for cleaning and preparing feather samples for analysis is detailed in Table 3.1.

In order to be successful, the comparative equilibration approach for SIA relies on the long term availability of keratin and organic tissue working standards because it is essential that there is rigid QA and comparability of results between laboratories to ensure accuracy of results. Although there are some primary, certified organic standards – e.g. IAEA-CH-7 which has been checked for isotopic homogeneity and has a recommended $\delta^{13}C$ value determined by international calibration – they do not contain exchangeable hydrogen and are not suitable reference material for calibrating δD in feathers. Three keratin standards have been prepared at the laboratories of Environment Canada, Saskatoon, Saskatchewan, Canada from cryogenically ground and isotopically homogenized chicken feathers (CFS), cow hoof (CHS) and bowhead whale baleen (BWB-II). They were calibrated according to Wassenaar and Hobson (2003). CFS was obtained from a single batch of 2 kg of chicken feathers; CHS was prepared from a hoof of a cow obtained from an abattoir and BWB-II was obtained from the University of Alaska. All of the keratins were solvent cleaned using methanol/chloroform, air-dried and cryogenically homogenized to about 0.5 kg. They were passed through <100 μm sieves and further homogenized to ensure isotopic homogeneity at the 100 μg level. These standards were sufficient to last for several years of use in a single laboratory. The results for δD were CFS, −147±5 %, (VMSOW), CHS, 187±2 % (VMSOW) and BWB-II, −108±4 % (VMSOW).

It is essential that isotopic working standards for δD in keratinous materials are produced to provide for the long-term needs of migration and ecological research. Such materials would need to be: (i) in sufficient quantities to allow their use by many different laboratories working on stable isotopes for up to 10 years; (ii) be certified for δD homogeneity to <100 μg and (iii) be certified for an isotopic range of at least 200 %. This need for standardization will become more important as studies on stable isotope for migration and other reasons increases to become a mainstream analytical tool.

3.2   Other Stable Isotopes for Migration Research

Stable oxygen isotopes $\delta^{18}O$ could also be used for migration research, especially as keratins have no exchangeable oxygen so there is no need for comparative equilibration procedures. The disadvantage is that the $\delta^{18}O$ range for tissues is small (approx. 15‰) and the analytical error for measurements are high so that precision regarding geospatial information is lost. Carbon ($\delta^{13}C$) and nitrogen ($\delta^{15}N$) isotopes can also be used in migration studies for local-spatial analysis, to refine populations or indicate habitat type.

3.2.1   Sampling Instructions for Water $\delta^{18}O$ and δ^2H – Rivers/ Lakes/Groundwater

Materials

Twenty five to sixty milliliter of HDPE (plastic) bottles with caps. Only a few ml of water is required for the analysis in the laboratory, but very small bottles are awkward for field staff to clearly write labels on so for logistical reasons use bottles of a capacity of 2 ml. Do not use glass bottles since they can shatter in transport. Evaporation and leakage during storage must be avoided. This can occur by mishandling or from loose caps.

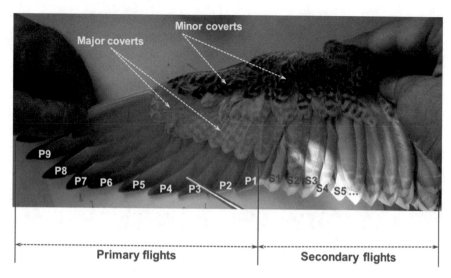

Fig. 3.1 Nomenclature of wing feathers of birds (Sample from the CRP D32030)

Method

Fill sample bottle with raw river, ground or lake water to overflowing. Cap tightly. Before sampling ensure the sample bottle is clearly and labeled with site or station, and dated using permanent ink. Ensure label does not smear. Do not freeze. Store at room temperature. No preservation is required. Samples should be shipped ground or air with adequate protection (box with foam chips, bubble wrap) and sent to the isotope laboratory of choice.

3.2.2   Sampling Instructions for Bird Feathers for Use in SIA

Materials

Steel laboratory tweezers.
Steel Scissors
Paper Envelope – Sealable paper envelopes are best to use for storage of bird feathers. Stored feathers should not be wet. Sealable plastic bags can be used as well.
Long wing feathers: Cut long feather from wing (Fig. 3.1) with pair of scissors (preferably P1 or P2 feathers).
Down feathers: Pluck down feathers (5–10) with tweezers from wing, or breast.
If possible check bird body for Blood keels (feathers in growth) (Fig. 3.2): Pluck blood keels (down feathers in growth) with tweezers from breast, or cut one large blood keel (large feather in growth) with pair of scissors from wing. Feathers

Fig. 3.2 Blood keel feathers in growth

should be dry. No visible infestation of feather inhabiting parasites. Put all feathers from one individual bird in the same envelope. Close, seal and label the envelope. Note which samples it contains (large feather, down feathers, blood keel(s); bird species, [if possible also: sex, growths status (juvenile, adult)], date and location of sampling, name of person sampling. The samples can be stored at room temperature. No preservation is required. They should be shipped by ground or air with adequate protection (bubble wrap envelope) and sent to the isotope laboratory of choice.

3.2.3   Sampling Wild Birds in the Field

A comprehensive guide and strategy for sampling wild birds is given by Whitworth et al. (2007) and that manual should be referred to in dealing with HPAI (http://www.fao.org/docrep/010/a1521e/a1521e00.htm). The following section highlights a few relevant points.

On arrival at the site, evaluate the extent of the mortality, if appropriate, including number of birds, species directly involved, other wildlife or domestic animals involved, and geographic range of mortality. This information should be recorded in a Sick or Dead Bird Sample Collection Log. In addition to preparing for animal sample collection, also consider collection of other environmental samples including water, soil, vegetation or other elements from which it might be possible to identify HPAI virus. If it is possible to get a GPS locations that characterize the extent of the die-off area, this is preferable to a general verbal description.

Wear the appropriate level of personal protective equipment, based on the situation being investigated. Try to minimize direct contact with dead birds and always keep animals away from the face, wear vinyl or latex gloves if handling a dead bird.

The best method for collecting a dead bird is to invert a plastic bag around your gloved hand and then surround the animal with the bag so that the bird is not directly touched. Seal the bag tightly (double bag if required for strength and cleanliness) and clearly and indelibly label the bag with an Animal Identification Number (which must match the number entered in the Sick or Dead Bird Sample Collection Log), species, date, time and location. If more than one species has been affected, collect several specimens of each for diagnosis. In general, carcasses of birds that have been dead for less than 24 h (fresh carcasses) are sufficiently adequate (moribund or viraemic birds are best) for diagnostic purposes. In colder climates, carcasses may last in relatively good condition for longer periods of time; in warm climates, carcasses will decompose faster.

When possible, fresh carcasses should be refrigerated (NOT frozen); a decomposing carcass is desiccated, bloated, green, foul smelling and has feathers that pull out easily. To increase diagnostic value, fresh carcasses must be transferred to the appropriate veterinary or pathology facility and examined as soon as possible. In field settings and/or far from appropriate diagnostic facilities, collect samples on site and place them in an ice chest or cooler. Keep carcasses away from refrigerators used for animal or human food.

3.2.4   Sampling Strategy for H5N1

For each affected species, select up to 3 birds that have most recently died (less than 24 h) or more if practical, up to 3 sick birds (suffering respiratory, neurologic or gastrointestinal disease or moribund) and up to 3 apparently healthy birds in direct contact with currently sick birds. If possible, also conduct a survey of other live birds that share the same habitat (cloacal swabs and/or tracheal swabs only). Priority should be given to birds that share wetlands with affected birds since the main mode of transmission of AI virus is probably faecal contamination of water, shores or banks. It is best to collect as many carcasses as possible and to place them in a central location for processing. Removal of dead birds from the site may also help prevent secondary contamination of scavengers or the environment. It is very important to complete the Sick or Dead Bird Sample Collection Log as carcasses are being collected and processed. If possible, try to collect and examine sick animals as well as newly dead birds, making sure that you have the appropriate permits to capture live samples. If there are too many dead to be able to individually bag and label, try to bag or examine well preserved animals that will be most useful for diagnostic purposes, and keep these separate from decomposing carcasses. If possible, transport carcasses (in sealed bags) in a space well separated from the occupants of the vehicle. If you are working in a remote area, you may have to perform field necropsies on site.

Recommended ornithological information to be collected during surveillance programmes or field assessment of wild bird mortality events.

All birds from which samples are taken should be identified to species. Where clearly distinguishable subspecies or discrete populations exist, as for some geese, this information should also be collected and reported. Age and sex should be recorded wherever possible. Close collaboration with ornithologists in the capture and sampling of live birds not only facilitates identification of birds but also gives the opportunity to collect additional information on the sampled live birds (such as weight, age, sex and condition), which are important to developing a better understanding of viral ecology and epidemiology. Standard protocols exist for the collection of such data through national ringing schemes [details of which are available for Europe, for example, via EURING (www.EURING.org)]. Recording individual ring numbers in the reporting spreadsheet provides a means of accessing these data for future analysis. To provide an audit of identification, it is highly desirable that a clear digital photograph is taken of each sampled bird (especially those found dead and/or not identified by ornithologists) and stored at least until confirmation of laboratory tests. In order to facilitate identification of bird species (which can sometime vary in quite minor plumage details, especially at certain times of the year), photographs should be taken according to the guidance given in Sect. 3.2. In the event of positive results, further examination of such photos can provide additional information on the age and sex of the bird, in addition to proving the identity of the species beyond doubt and thus allowing the case to be correctly put into context. To facilitate this, each individual bird should be given a code that is used on the cloacal and oropharyngeal swabs taken, and this code should be on a piece of card that is visible in each photograph taken.

Especially related to sampling in the vicinity of outbreaks, it is desirable to collect a range of contextual information so as to better understand the viral epidemiology of H5N1 HPAI in wild bird populations. Such information should include:

- date of sampling, clear locational and descriptive data about the catching site, ideally GPS coordinates, including habitat description (e.g., lake, river, village pond, fish farm, etc.) and distance to human settlement, agricultural land, and poultry farms; it may also be useful to include details about the season and relate this information to the natural behaviour/cycle of the affected birds, e.g., moulting, premigration, during migration, etc.;
- record of the numbers of each species of other live birds in the sampling area that were not sampled;
- if available, records of bird movements (arrivals/departures) that occurred at the sampling site prior to the sampling;
- assessment of the numbers of each species of live bird in the sampling area that were not sampled but that were showing signs of ill health; and
- given that birds of some species (such as Mallards-Anas platyrhynchos) can occur either as wild birds that are able to move between sites or occur in a feral state, habituated to foods provided by humans, distinguishing between these categories would be useful. Sometimes the presence of unusual plumage patterns, indicating domestication, is useful in this respect.

Taking photographs of dead birds for identification purposes

Different bird species are identified by differing characteristics, so it is difficult to provide universal guidance applicable in all situations. However, the following is a minimum standard that should be followed. All wild birds collected for analysis for HPAI should have digital photographs taken as soon as possible after collection. The bird should fully fill the photograph and wherever possible include a ruler or other scale measure. Each photograph should be taken at the highest resolution possible and if the camera has a 'date stamp' feature then this should be enabled so that the image is saved with a time reference – this may help verify the sequence of images taken at a site on a day. Images should be downloaded to a computer as soon as possible and information about location and date added to the file properties.

Photographs should be taken of:

- The whole bird, dorsal side, with one wing stretched out and tail spread and visible;
- The head in profile clearly showing the beak;
- Close-up photos of the tips of wing feathers, as these can often determine whether the bird is an adult or a juvenile (bird in its first year); and
- Ideally photographs of both dorsal and ventral views of the bird, as photos of the upper and under surfaces of the wing and spread tail will facilitate aging and sexing of birds

Any ventral photographs should show the legs and feet (since leg color is often an important species diagnostic). If any rings (metal or plastic) are present on the legs, these should be photographed *in situ* as well as recording ring details. Any conspicuous markings/patterns should also be photographed.

At certain times of the year, such as late summer (July–August in the northern hemisphere) many waterbirds, especially ducks and geese, undergo moult and can be especially difficult to identify by non-specialists. At such times clear photographs are especially important to aid identification of (duck) carcasses. The patch of colour on the open wing (called the "speculum") is often especially useful. The identification of young gulls at any time of the year is also difficult and typically they will also need to be photographed and identified by specialists.

Photographs should be retained, linked to an individual specimen, at least until laboratory tests are returned as negative for avian influenza. A unique code or reference number that is the same as the code or reference number of any samples taken from the birds should be visible in each photograph so as to link samples and photographs. Photographs can be used immediately if identification of the species of bird is in any doubt, and for subsequent checking of the identification if necessary.

Correction to: Stable Isotopes to Trace Migratory Birds and to Identify Harmful Diseases

G. J. Viljoen, A. G. Luckins, and I. Naletoski

Correction to:
G.J. Viljoen et al., *Stable Isotopes to Trace Migratory Birds and to Identify Harmful Diseases*, https://doi.org/10.1007/978-3-319-28298-5

The book was inadvertently published with materials that were reproduced or modified without permission, attribution, or citation by using original scientific materials taken from the textbook "Tracking animal migration with stable isotopes (Vol. 1): Hobson, K. A., & Wassenaar, L. I., (2008) by Academic Press, in chapters 2 and 3.

The book has now been updated by including an Acknowledgement and Correction to the specific contributions of the chapter authors in the original 2008 book edited by K. A. Hobson & L. I. Wassenaar.

Following are the specific instances of materials that were reproduced without permission and hereby are listed with the correct attribution:

Revised Table Headings

Table 2.1 Extrinsic markers for tracking animal migration. Reproduced with Permission from Hobson, K. A., & Norris, D. R. (2008). Animal migration: a context for using new techniques and approaches. *Terrestrial Ecology, 2*, 1–19.

The updated online version of the chapters 2 and 3 can be found at
https://doi.org/10.1007/978-3-319-28298-5_2
https://doi.org/10.1007/978-3-319-28298-5_3

The updated version of this book can be found at
https://doi.org/10.1007/978-3-319-28298-5

© IAEA 2022
G.J. Viljoen et al., *Stable Isotopes to Trace Migratory Birds and to Identify Harmful Diseases*, https://doi.org/10.1007/978-3-319-28298-5_4

Table 2.2 Intrinsic markers for tracking animal migration stable isotopes. Reproduced with Permission from Hobson, K. A., & Norris, D. R. (2008). Animal migration: a context for using new techniques and approaches. *Terrestrial Ecology, 2*, 1–19.

Table 2.3 Dry weight % abundance of light stable isotope ratios of interest in determining migratory connectivity in tissues. Reproduced with Permission from Wassenaar, L. I. (2008). An introduction to light stable isotopes for use in terrestrial animal migration studies. *Terrestrial Ecology, 2*, 21–44.

Table 2.5 Methods for assigning animals of unknown origin using stable isotopes. Reproduced with Permission from Wunder, M. B., & Norris, D. R. (2008a). Analysis and design for isotope-based studies of migratory animals. *Terrestrial Ecology, 2*, 107–128.

Table 3.1 Procedure for preparing feather samples for δD stable isotope analysis. Reproduced with Permission from Wassenaar, L. I. (2008). An introduction to light stable isotopes for use in terrestrial animal migration studies. *Terrestrial Ecology, 2*, 21–44.

The acknowledgements have been updated to reflect the contributions of K.A. Hobson and L.I. Wassenaar to the project, IAEA CRP Code: D32030, as follows:

Acknowledgements:

This guide was developed with a substantial support of the team involved in the development and implementation of the IAEA CRP: "Use of Stable Isotopes to Trace Bird Migrations and Molecular Nuclear Techniques to Investigate the Epidemiology and Ecology of the Highly Pathogenic Avian Influenza" (Code: D32030).

A second edition of *Stable Isotopes to Trace Migratory Birds and to Identify Harmful Diseases – An Introductory Guide* is in preparation and will supersede the original published in 2016.

References:

Tracking Animal Migration with Stable Isotopes, Edited by: Keith A. Hobson, Leonard I. Wassenaar, First Edition 2008 (ISBN: 978-0-12-373867-7, ISSN: 1936-7961).

References

Bearhop, S., S. Waldron, S.C. Votier, and R.W. Furness. 2002. Factors that influence assimilation rates and fractionation of nitrogen and carbon stable isotopes in avian blood and feathers. *Physiological and Biochemical Zoology* 75: 451–458.

Bowen, G.J., and J. Revenaugh. 2003. Interpolating the isotopic composition of modern meteoric precipitation. *Water Resources Research* 39(10): 1299. doi:10.129/2003WR002086.

Bowen, G.J., and J.B. West. 2008. *Isotope landscapes for terrestrial migration research: Tracking animal migration with stable isotopes*, 79–106. New York: Academic.

Bowen, G.J., and B. Wilkinson. 2002. Spatial distribution of $\delta^{18}O$ in meteoric precipitation. *Geology* 30(4): 315–318.

Bowen, G.J., L.I. Wassenaar, and K.A. Hobson. 2005. Global application of stable hydrogen and oxygen isotopes to wildlife forensics. *Oecologia* 143: 337–348.

Carleton, S., and C. Martinez del Rio. 2005. The effect of cold-induced increased metabolic rate on the rate of ^{13}C and ^{15}N incorporation in house sparrows (*Passer domesticus*). *Oecologia* 144: 226–232.

Chang, Yuan-Mou, K.A. Hatch, Tzung-Su Ding, D.L. Eggett, Hsiao-Wei Yuan, and B.L. Roeder. 2008. Using stable isotopes to unravel and predict the origins of great cormorants (*Phalacrocorax carbo sinensis*) overwintering at Kinmen. *Rapid Communications in Mass Spectrometry* 22: 1235–1244.

Cheung, Peter P., Y.H. Connie Leung, Chun-Kin Chow, Chi-Fung Ng, Chun-Lok Tsang, Yu.-On. Wu, Siu-Kit Ma, Sin-Fun Sia, Yi. Guan, and J.S.M. Peiris. 2009. Identifying the species origin of faecal droppings used for avian influenza virus surveillance in wild birds. *Journal of Clinical Virology* 46: 90–93.

Clark, I.D., and P. Fritz. 1997. *Environmental isotopes in hydrogeology*. New York: Lewis Publishers, 328 pp.

Darling, W.G., and A.H. Bath. 1988. A stable isotope study of recharge processes in the English chalk. *Journal of Hydrology* 101: 31–46.

de'Ath, Glenn, and Katharina E. Fabricius. 2000. Classification and regression trees: A powerful yet simple technique for ecological data analysis. *Ecology* 81: 3178–3192.

The original version of this chapter was revised: We have to insert the references for author Hobson and Norris (2008), Wassenaar (2008), Wunder and Norris (2008a) in References list and also fix the "b" for the author Wunder and Norris (2008b) on page 48 has been corrected now. The correction to this chapter is available at https://doi.org/10.1007/978-3-319-28298-5_4

Dovas, C.I., M. Papanastassopoulou, M.P. Georgiadis, E. Chatzinasiou, V.I. Maliogka, and G.K. Georgiades. 2010. Detection and quantification of infectious avian influenza A (H5N1) virus in environmental water by using real-time reverse transcription-PCR. *Applied and Environmental Microbiology* 76: 2165–2174.

Evans Ogden, L.J., K.A. Hobson, and D.B. Lank. 2004. Blood isotopic (d13C and d15N) turnover and diet-tissue fractionation factors in captive Dunlin (*Calidris alpina pacifica*). *Auk* 121: 170–177.

FAO-Global Animal Disease Information System (EMPRES-i). 2015. http://empres-i.fao.org/eipws3g/. Last accessed May 2015.

Greenberg, R., P.P. Marra, and M.J. Wooler. 2007. Stable-isotope (C, N, H) analyses help locate the winter range of the coastal plain swamp sparrow (*Melospiza georgiana nigrescens*). *Auk* 124: 1137–1148.

Hebert, Craig E., and Leonard I. Wassenaar. 2005. Feather stable isotopes in western North American waterfowl: Spatial patterns, underlying factors, and management applications. *Wildlife Society Bulletin* 33: 92–102.

Hiroyoshi, H., and J.P. Pierre. 2005. Satellite tracking and avian conservation in Asia. *Landscape Ecological Engineering* 1: 33–42.

Hobson, K.A. 2002. Incredible journeys. *Science* 295: 981.

Hobson, K.A. 2008. Isotopic methods to track animal movements. In *Tracking animal migration with stable isotopes*, ed. K.A. Hobson and L.I. Wassenaar. Oxford: Elsevier.

Hobson, K.A., and R.G. Clark. 1992. Assessing avian diets using stable isotopes. 1. Turnover of C-13 in tissues. *Condor* 94: 181–188.

Hobson, K.A., and R.G. Clark. 1993. Turnover of ^{13}C in cellular and plasma fractions of blood: Implications for non-destructive sampling in avian dietary studies. *Auk* 110: 638–641.

Hobson, Keith A., and Leonard I. Wassenaar. 1997. References linking breeding and wintering grounds of neotropical migrant songbirds using stable hydrogen isotopic analysis of feathers. *Oecologia* 109: 142–148.

Hobson, K.A., and D.R. Norris. 2008. Animal migration: a context for using new techniques and approaches. *Terrestrial Ecology* 2: 1–19.

Hobson, K.A., L.I. Wassenaar, and E. Bayne. 2004. Using isotopic variance to detect long-distance dispersal and philopatry in birds: An example with ovenbirds and American redstarts. *The Condor* 106: 732–743.

Hobson, Keith A., Steven Van Wilgenburg, Leonard I. Wassenaar, Helen Hands, William P. Johnson, Mike O'Meilia, and Philip Taylor. 2006. Using stable hydrogen isotope analysis of feathers to delineate origins of harvested sandhill cranes in the central flyway of North America. *Waterbirds* 29: 137–147.

Hobson, Keith A., Steve Van Wilgenburg, Leonard I. Wassenaar, Frank Moore, and Jeffrey Farrington. 2007. Estimating origins of three species of neotropical migrant songbirds at a gulf coast stopover site: Combining stable isotope and GIS tools. *Condor* 109: 256–267.

Hobson, Keith A., Steven L. Van Wilgenburg, Keith Larson, and Leonard I. Wassenaar. 2009. A feather hydrogen isoscape for Mexico. *Journal of Geochemical Exploration* 102: 63–70.

IAEA. 2001. *Isotope hydrology information system*. The ISOHIS, database. Source:http://www.isohis.iaea.org.

IAEA. 2006. *Global Network of Isotopes in Precipitation (GNIP)*. International Atomic Energy, Agency, Vienna, Austria. Source:http://www-naweb.iaea.org/napc/ih/IHS_resources_gnip.html.

IAEA. 2007. *Atlas of isotope hydrology – Africa*. Vienna: International Atomic Energy Agency.

Kelly, Jeffrey F., Viorel Atudorei, Zachary D. Sharp, and Deborah M. Finch. 2002. Insights into Wilson's Warbler migration from analyses of hydrogen stable-isotope ratios. *Oecologia* 130: 216–221.

Kelly, Jeffrey F., Kristen C. Ruegg, and Thomas B. Smith. 2005. Combining isotopic and genetic markers to identify breeding origins of migrant birds. *Ecological Applications* 15: 1487–1494.

Khalenkov, A., W.G. Laver, and R.G. Webster. 2008. Detection and isolation of H5N1 influenza virus from large volumes of natural water. *Journal of Virological Methods* 149: 180–183.

Lee, Dong-Hun, Hyun-Jeong Lee, Yu.-Na. Lee, Youn-Jeong Lee, Ok.-Mi. Jeong, Hyun-Mi Kang, Min-Chul Kim, Ji-Sun Kwon, Jun-Hun Kwon, Joong-Bok Lee, Seung-Yong Park, In-Soo Choi, and Chang-Seon Song. 2010. Application of DNA barcoding technique in avian influenza virus surveillance of wild bird habitats in Korea and Mongolia. *Avian Diseases* 54: 677–681.

McKechnie, A.E. 2004. Stable isotopes: Powerful new tools for animal ecologists. *South African Journal of Science* 100: 131–134.

Morrison, R.I.G., and K.A. Hobson. 2004. Use of body stores in shorebirds after arrival on high-arctic breeding grounds. *Auk* 121: 333–344.

Norris, D.R., P.P. Marra, G.J. Bowen, L.M. Ratcliffe, J.A. Royle, and T.K. Kyser. 2006. Migratory connectivity of a widely distributed songbird, the American Redstart (Setophaga ruticilla). *Ornithological Monographs* 61: 14–28.

Oppel, S., and A.N. Powell. 2010. Carbon isotope turnover in blood as a measure of arrival time in migratory birds using isotopically distinct environments. *Journal of Ornithology* 151: 123–131.

Pérez, G.E., K.A. Hobson, E.J. Garde, and M. Gilbert. 2010. Deuterium (δD) in feathers of Mongolian waterbirds uncovers migratory movements. *Waterbirds* 33: 438–443.

Peterson, B.J., and B. Fry. 1987. Stable isotopes in ecosystem studies. *Annual Review of Ecology and Systematics* 18: 293–320.

Phillips, D.L., and P.M. Eldridge. 2006. Estimating the timing of diet shifts using stable isotopes. *Oecologia* 147: 195–203.

Prosser, D.J., P. Cui, J.Y. Takekawa, M. Tang, Y. Hou, Bridget M. Collins, Baoping Yan, Nichola J. Hill, Tianxian Li, Li. Yongdong, Fumin Guo Lei, Shan Xing, Zhi He, Yuanchun Yubang, D.A. Rocque, Merav Ben-David, P. Barry Ronald, and Kevin Winker. 2006. Assigning birds to wintering and breeding grounds using stable isotopes: Lessons from two feather generations among three intercontinental migrants. *Journal of Ornithology* 147: 395–404.

Rozanski, K. 1985. Deuterium and ^{18}O in European groundwaters – Links to atmospheric circulation in the past. *Chemical Geology* 52: 349–363.

Smith, R.B., E.C. Greiner, and B.O. Wolf. 2004. Migratory movements of sharp-shinned hawks (*Accipiter striatus*) captured in New Mexico in relation to prevalence, intensity, and biogeography of avian hematozoa. *Auk* 121: 837–846.

Stutchbury, B.J.M., S.A. Tarof, T. Done, E. Gow, P.M. Kramer, J. Tautin, J.W. Fox, and V. Afanasyev. 2009. Tracking long-distance songbird migration using geolocators. *Science* 323: 896.

Szé'p, T., K.A. Hobson, J. Vallner, S.E. Piper, B. Kova´cs, D.Z. Szabo, and A.P. Møller. 2008. Comparison of trace element and stable isotope approaches to the study of migratory connectivity: An example using two hirundine species breeding in Europe and wintering in Africa. *Journal of Ornithology* 150: 621–636.

Wassenaar, L.I. 2008. An introduction to light stable isotopes for use in terrestrial animal migration studies. *Terrestrial Ecology* 2: 21–44.

Wassenaar, L.I., and K.A. Hobson. 2000. Stable carbon and hydrogen isotope ratios reveal breeding origins of red-winged blackbirds. *Ecological Applications* 10: 911–916.

Wassenaar, L.I., and K.A. Hobson. 2003. Comparative equilibration and online technique for determination of non-exchangeable hydrogen of keratins for use in animal migration studies. *Isotopes in Environmental and Health Studies* 39: 211–217.

Wassenaar, L.I., and K.A. Hobson. 2006. Stable-hydrogen isotope heterogeneity in keratinous materials: Mass spectrometry and migratory wildlife tissue subsampling strategies. *Rapid Communications in Mass Spectrometry* 20: 2505–2510.

Whitworth, D., S. Newman, T. Mundkur, and P. Harris. 2007. Wild birds and avian influenza, an introduction to applied field research and disease sampling techniques, *FAO Animal Production and Health, Manual No. 5, Food and Agriculture Organization of The United Nations, Rome, 2007*, http://www.fao.org/docrep/010/a1521e/a1521e00.htm.

Wunder, M.B. 2010. Using isoscapes to model probability surfaces for determining geographical origins. In *Isoscapes: Understanding movements, pattern, and process on earth through isotope mapping*, ed. J.B. West, G.J. Bowen, T.E. Dawson, and K.P. Tu, 251–272. New York: Springer.

Wunder, M.B., and D.R. Norris. 2008a. Analysis and design for isotope-based studies of migratory animals. *Terrestrial Ecology* 2: 107–128.

Wunder, Michael B., and D. Ryan Norris. 2008b. Improved estimates of certainty in stable-isotope-based methods for tracking migratory animals. *Ecological Applications* 18: 549–559.

Yamaguchi, Noriyuki, Jerry Hupp, Hiroyoshi Higuchi, Paul Flint, and John Pearce. 2010. Satellite-tracking of northern pintail during outbreaks of the H5N1 virus in Japan. *Implications for Virus Spread Ibis* 152: 262–271.

Yohannes, E., K.A. Hobson, and D.J. Pearson. 2007. Feather stable-isotope profiles reveal stop-over habitat selection and site fidelity in nine migratory species moving through sub-Saharan Africa. *Journal of Avian Biology* 38: 347–355.

Zhou, Douglas, C. David, William M. Perry, and Scott H. Newman. 2011. Wild bird migration across the Qinghai-Tibetan plateau: A transmission route for highly pathogenic H5N1. *PLoS One* 6(3): e17622. doi:10.1371/journal.pone.0017622.

Index

© IAEA 2016
G.J. Viljoen et al., *Stable Isotopes to Trace Migratory Birds and to Identify
Harmful Diseases*, DOI 10.1007/978-3-319-28298-5